Learning Materials in Biosciences

Learning Materials in Biosciences textbooks compactly and concisely discuss a specific biological, biomedical, biochemical, bioengineering or cell biologic topic. The textbooks in this series are based on lectures for upper-level undergraduates, master's and graduate students, presented and written by authoritative figures in the field at leading universities around the globe.

The titles are organized to guide the reader to a deeper understanding of the concepts covered.

Each textbook provides readers with fundamental insights into the subject and prepares them to independently pursue further thinking and research on the topic. Colored figures, step-by-step protocols and take-home messages offer an accessible approach to learning and understanding.

In addition to being designed to benefit students, Learning Materials textbooks represent a valuable tool for lecturers and teachers, helping them to prepare their own respective coursework.

More information about this series at https://link.springer.com/bookseries/15430

Neena Grover

Editor

Fundamentals of RNA Structure and Function

 Springer

Editor
Neena Grover
Department of Chemistry and Biochemistry
Colorado College
Colorado, CO
USA

ISSN 2509-6125   ISSN 2509-6133   (electronic)
Learning Materials in Biosciences
ISBN 978-3-030-90213-1   ISBN 978-3-030-90214-8   (eBook)
https://doi.org/10.1007/978-3-030-90214-8

This Springer imprint is published by the registered company Springer Nature Switzerland AG
The registered company address is: Gewerbestrasse 11, 6330 Cham, Switzerland

To My Family

Preface

Our journey into the "RNA World" begins with the discovery of catalytic functions in RNA; it encompasses the exploration of the hypothetical early world where RNA are the first genetic molecules, and brings us into the current world of exciting new discoveries in RNA's structure and function. The last three decades provide a glimpse into the large regulatory networks formed by RNA. Simultaneously, the technological revolutions in RNA crystallography, cryo-electron microscopy (EM), and genomic analyses add a new depth to our molecular view of the biological systems.

RNA biochemistry reveals the beauty and complexity of structures and their functions. The topics I include here are those that should add new dimensions and depth into the undergraduate biochemistry and molecular biology curriculum.

This book is a guide for you if you are interested in entering into the RNA world. I use the book chapters to provide background information to students before we explore the topics further. *This book is a resource to guide further investigations; it is not a textbook nor a collection of review articles.* It is meant for an upper-level undergraduate course based in literature. For reference, students with one biochemistry course (proteins and metabolism) and some introductory cellular and molecular biology courses do well in my RNA course.

When I teach my RNA course, it is difficult to find RNA review articles at the Goldilocks level of detail. So, my students and I started writing these essays *for our own use.* This is the story of the birth of this book. The students are listed as coauthors on the chapters.

I have transformed the original writings into chapters in the hope that these will be useful for those interested in learning more about RNA.

The referencing in the book is light, per book guidelines. My RNA classes start with techniques in RNA, which I have not included here. The figures are meant to make the material visual. Do read the figure captions as additional information is included here. There are beautiful figures available in the literature to supplement those included here. I am fully responsible for the final product.

Most undergraduate professors with sufficient biochemical knowledge should find the book usable. If you are new to the RNA field, read Chaps. 1–3 to acquire the vocabulary of RNA structures and catalysis; you may read subsequent chapters in any order. I am including the protein data bank (PDB) file numbers in the figure legends, to allow 3D viewing of molecular structures, a must in my mind. I use PyMol but any viewer should work.

I predominantly use RNA as a singular and plural noun (as is often seen in the literature) with the assumption that A stands for acid or acids; there is no set convention.

Chapter 1 reviews the composition of nucleic acids. A decent understanding of noncanonical base pairs is necessary to understand RNA structures and functions.

Chapter 2 introduces RNA structures. Students generally like to keep this chapter on hand as a reference for the structural vocabulary. I encourage students to research details of new RNA structures that they encounter in their research topics.

Chapter 3 is on small catalytic RNA which are model systems to understand the principles of phosphodiester bond cleavage and ligation. RNA structures are modular; therefore, the structural patterns seen in these small RNA repeat in the larger RNA.

Chapter 4 introduces the complex and dynamic nature of the spliceosome assembly using cryo-EM images collected in the last decade. These images assist in validating and updating the structural models proposed for the various steps of assembly and associated reactions. I use small group for individual steps before bringing it all together as a class. It is a complex topic that is worth the time investment.

Chapter 5 is on RNA sequencing technology. Sara Hanson is an expert in this area. She and her students provide an introduction to genomics and its applications that show up in various chapters.

Chapters 6–9 are on mRNA and its regulation. The *in vitro* and *in vivo* analyses show a structural code within the sequence code of mRNA. I include a section on mRNA vaccines to encourage students to participate in the dissemination of science. For example, my nucleic acid biochemistry class organized and participated in a panel on Race, Racism and Vaccines.

Chapter 7 is about the riboswitches and their role in gene expression.

Chapter 8 is on the new and exciting area of non-coding RNA-based cell regulation. I focus on a small area of microRNA biogenesis. RNA interference is changing our understanding of regulation networks in the cells and our practice of biology (e.g., genes silencing vs gene knockouts).

Chapter 9 is on the fascinating biochemistry of CRISPR-Cas systems. CRISPR, like mRNA vaccines, is in the public vernacular. I discuss the science of CRISPR along with the associated ethical implications. I have provided a short compilation of ethical issues that are currently being discussed.

Chapter 10 is a short introduction to the steps involved in transcription, a complex field that is very well researched in both bacteria and in eukarya.

The stories end here for now.

I have tried to move the molecular conversations away from the aggressive, competitive, industrial language that often permeates science, perhaps reflecting the current state of science. Using a more inclusive language may also help us to move the culture in the direction of cooperation and inclusion and lessen the focus on individual scientists. If our research is to benefit all members of the society then we have to examine the structures of science.

One often hears about the problem of diversity and equity as one of lack of students of color in our classrooms. It is imperative that all our classrooms be more diverse. We need to teach with broader perspectives. The micro- and macro-aggressions in our classrooms and research laboratories are rampant and unacceptable. Women are routinely exploited and silenced. People of color are routinely marginalized.

Currently, most principal investigators in the USA, including this field, are white. The four underrepresented groups of minorities in the USA are largely missing from our community and from the larger intellectual enterprise. We also have to do better by the foreign graduate students and postdoctoral researchers (many are people of color) who are performing groundbreaking research every day. The racism and xenophobia in science is impacting our intellectual output at a time when we face grave challenges.

If we were to be inclusive and diverse, it would impact all our choices from our model systems to our leadership. First, we have to first stop comparing people to leaks and problems to pipelines. We need to humanize our problems and work from a place of empathy. We have the intellectual capacity to do better at every level! We can do better! We must!

There are medical implications to the RNA research presented here. We have to include conversations on ethics when discussing these topics. There are materials available to build skills in leading these "murky" conversations. Our social sciences or philosophy colleagues can also help us to lead these conversations. I collaborated with a social scientist, a faculty member in Feminist and Gender Studies, and her Black Feminist Theory class to run a Race, Racism and Vaccine panel.

We have a lot to do in science; let's include all those who are willing to participate in this journey. We have an obligation to better science and be better scientists; let's press on.

I am no longer accepting the things that I cannot change. . . I am changing the things I cannot accept.—Angela Davis

Colorado, CO, USA   Neena Grover

Acknowledgments

My nucleic acid research started in Holden Thorp's laboratory with DNA and metal ions. My education in RNA started at Boulder where my mentors, colleagues, and friends in Olke Uhlenbeck's and Tom Cech's laboratories were extraordinary teachers. Holden gave me the confidence to believe in myself. Liz Theil's mentoring led me to RNA. I learned RNA thermodynamics from the master himself, the famous Doug Turner, with additional help from David Matthews. Eric Westhof graciously allowed me to do my sabbaticals in his laboratory where I worked with Pascal Auffinger. Pascal has taught me about ions in structures. He is a wonderful colleague and friend. I have learned much from the many RNA scientists at CNRS, Strasbourg. My RNA friends and my students continue to inspire me to learn more.

Jenn Garcia and Sara Hanson have generously contributed to the book. Fred LaRiviere's support and friendship as a fellow RNA person have been invaluable. Many RNA friends provided last-minute feedback on different chapters. No one has done more than Quentin Vicens who read several chapters and was always supportive. Pascal Auffinger, Chrysa Latrick, and Maddie Sherlock all provided quick and thoughtful feedback. Many RNA colleagues have participated in my classes, including Quentin Vicens, Marino Resendiz, and David Booth. Kathy Guiffre has provided insights on the ethics section. Nate Bower and Harold Jones have continued to provide their editorial help even after their retirement. My biochemistry colleagues Peggy Daugherty and Annelise Gorensek-Benitz have generously provided feedback and encouragement. I am grateful to Mark Wilson, and all biology colleagues, for their feedback and collegiality. All my departmental colleagues are generous in their support and make it easy to come to work. Murphy Brasuel, Eli Fahrenkrug, and Jessica Kisunzu have particularly helped me with keeping teaching and DEI work at the forefront; they have been generous with their time. The undergraduate students at the Colorado College are amazing. They remind me to have high expectations! A special thank you to my research students who have worked with me over the years and have become colleagues and friends. Quinn Eaheart and Piper Catlin gave me their feedback on the early chapters, along with the RNA and nucleic acid course students.

None of this is possible without my family. My husband Gerald, and our children Kirby and Vivian, provide endless joy and love. The kids are always happy to see me, especially without my computer! My world would come to a standstill without Gerald's constant support. I can always count on love, support, and encouragement from my mom, my sister Mala (and her family), my brother Aseem (and his family), and dad Kirby II. My (late) dad always believed in me. The inspirational women in the family—my mom, my sister, my (late) grandmothers, and (late) mother-in-law Mary Dean—are (were) on the leading edge of thought in their areas!

Thanks especially to Alex Vargo, Ann Rule, and the California group for your friendship. Lexi, Ali, Ellie, and many others who are part of our extended family. The children's center teachers work hard to keep the kids safe; they often protect my (writing) time.

Amrei Strehl, my editor at Springer, is always gracious and kind. She answered my endless questions. Siva Ravanan worked to make the entire production process seamless. And thank you also to Beatrice Menz who believed in the possibility of this book before RNA was in the news, and Amrei for continuing that support over years. And thank you to the countless people who are working behind the scenes and don't get the recognition they deserve.

Contents

Contributors

Anjali Desai Department of Chemistry and Biochemistry, Colorado College, Colorado Springs, CO, USA

Chris Dickinson Department of Chemistry and Biochemistry, Colorado College, Colorado Springs, CO, USA

Jennifer F. Garcia Department of Molecular and Cellular Biology, Colorado College, Colorado Springs, CO, USA

Neena Grover Department of Chemistry and Biochemistry, Colorado College, Colorado Springs, CO, USA

Sara J. Hanson Department of Molecular Biology, Colorado College, Colorado Springs, CO, USA

Jake Heiser Department of Chemistry and Biochemistry, Colorado College, Colorado Springs, CO, USA

Hallie Hintz Department of Chemistry and Biochemistry, Colorado College, Colorado Springs, CO, USA

Cole Josefchak Department of Chemistry and Biochemistry, Colorado College, Colorado Springs, CO, USA

Evan Leslie Department of Chemistry and Biochemistry, Colorado College, Colorado Springs, CO, USA

Hayden Low Department of Molecular and Cellular Biology, Colorado College, Colorado Springs, CO, USA

Darryl McCaskill Department of Chemistry and Biochemistry, Colorado College, Colorado Springs, CO, USA

Jia A. Mei Department of Molecular Biology, Colorado College, Colorado Springs, CO, USA

Ethan Moore Department of Chemistry and Biochemistry, Colorado College, Colorado Springs, CO, USA

Natalie L. Nicholls Department of Chemistry and Biochemistry, Colorado College, Colorado Springs, CO, USA

Juliana C. Olliff Department of Molecular Biology, Colorado College, Colorado Springs, CO, USA

Human Biology Division, Fred Hutchinson Cancer Research Center, Seattle, WA, USA

Julia Poje Department of Chemistry and Biochemistry, Colorado College, Colorado Springs, CO, USA

Kathryn Reichard Department of Chemistry and Biochemistry, Colorado College, Colorado Springs, CO, USA

Sriya Sharma Department of Chemistry and Biochemistry, Colorado College, Colorado Springs, CO, USA

Kristie M. Shirley Department of Molecular Biology, Colorado College, Colorado Springs, CO, USA

Department of Chemistry and Biochemistry, Colorado College, Colorado Springs, CO, USA

Evgenia Shishkova Department of Chemistry and Biochemistry, Colorado College, Colorado Springs, CO, USA

Jordyn M. Wilcox Department of Molecular and Cellular Biology, Colorado College, Colorado Springs, CO, USA

Rachel Wilson Department of Chemistry and Biochemistry, Colorado College, Colorado Springs, CO, USA

Andy J. Wowor Department of Chemistry and Biochemistry, Colorado College, Colorado Springs, CO, USA

RNA: Composition and Base Pairing

1

Evan Leslie and Neena Grover

Contents

Keywords

RNA sugars · RNA phosphates · RNA nucleobases · RNA base pairs · RNA noncanonical base pairs · RNA backbone modifications

---

E. Leslie · N. Grover (✉)
Department of Chemistry and Biochemistry, Colorado College, Colorado Springs, CO, USA
e-mail: ngrover@ColoradoCollege.edu

© Springer Nature Switzerland AG 2022
N. Grover (ed.), *Fundamentals of RNA Structure and Function*, Learning Materials in Biosciences, https://doi.org/10.1007/978-3-030-90214-8_1

What You Will Learn

The sugar–phosphate backbone of nucleic acids binds to the heterocyclic, aromatic, nitrogenous bases. In this chapter, we will discuss sugar puckers, phosphodiester linkages, and base pairs. We will learn about noncanonical base pairs that are abundant in RNA structures. Concepts related to the nomenclature, hydrogen bonding, pKa, and isostericity of base pairs are a key to understanding the complex structures and functions in RNA.

Learning Objectives

After completing this chapter, the students should be able to:

- Draw the ribose and deoxyribose sugars in their most stable forms.
- Draw the phosphodiester linkage with the nucleobase connected. Number all the positions on the sugar and the base.
- Recognize hydrogen bond donors and acceptors and draw base pairs between any two bases.
- Recognize the nomenclature of noncanonical base pairs.
- Explain the concept of isosteres and predict how this might effect structures.
- Identify the potential impact of modifications and protonation on hydrogen bonding between bases.

1.1   Introduction

Nucleic acids were isolated in the late 1800s from the cell nuclei and were initially called nuclein. They were found to contain carbon, hydrogen, oxygen, nitrogen, and phosphorous. Despite evidence, our bias toward the complexity and importance of proteins made it difficult to convince ourselves that nucleic acids were the carriers of genetic information; it took until the 1940s before this debate was settled. Even though nucleic acids were first found in the nucleus, RNA molecules are mostly found in the cytoplasm. RNA are also secreted outside the cell and were recently found on the cell's surface bound to sugars, just like glycoproteins and glycolipids, and may have similar functions. We have learned much about nucleic acid since their discovery in the 1820s, yet a great deal remains unknown. An open mind and the right experiment might just lead you to the next big discovery! So let's get to it.

Nucleic acids are polymers made up of repeating units of sugars that are connected via phosphate molecules, with each sugar bound to one nucleobase. The nucleobases are connected to the sugar–phosphate backbone in a particular order. The sequence of nucleobases is the primary structure of a nucleic acid. In this chapter, we will examine each component and discuss the compositional differences between RNA and DNA.

1.2 Sugar

Nucleic acids utilize a five-carbon sugar, either deoxyribose or ribose, to build their polymeric structure (Fig. 1.1). The sugars of RNA and DNA exist in a ring form. The ring structures have angle and eclipsing strain. The angle strain in a ring is minimal when the ring carbons are close to being tetrahedral; this is the case for ring sizes of five to seven.

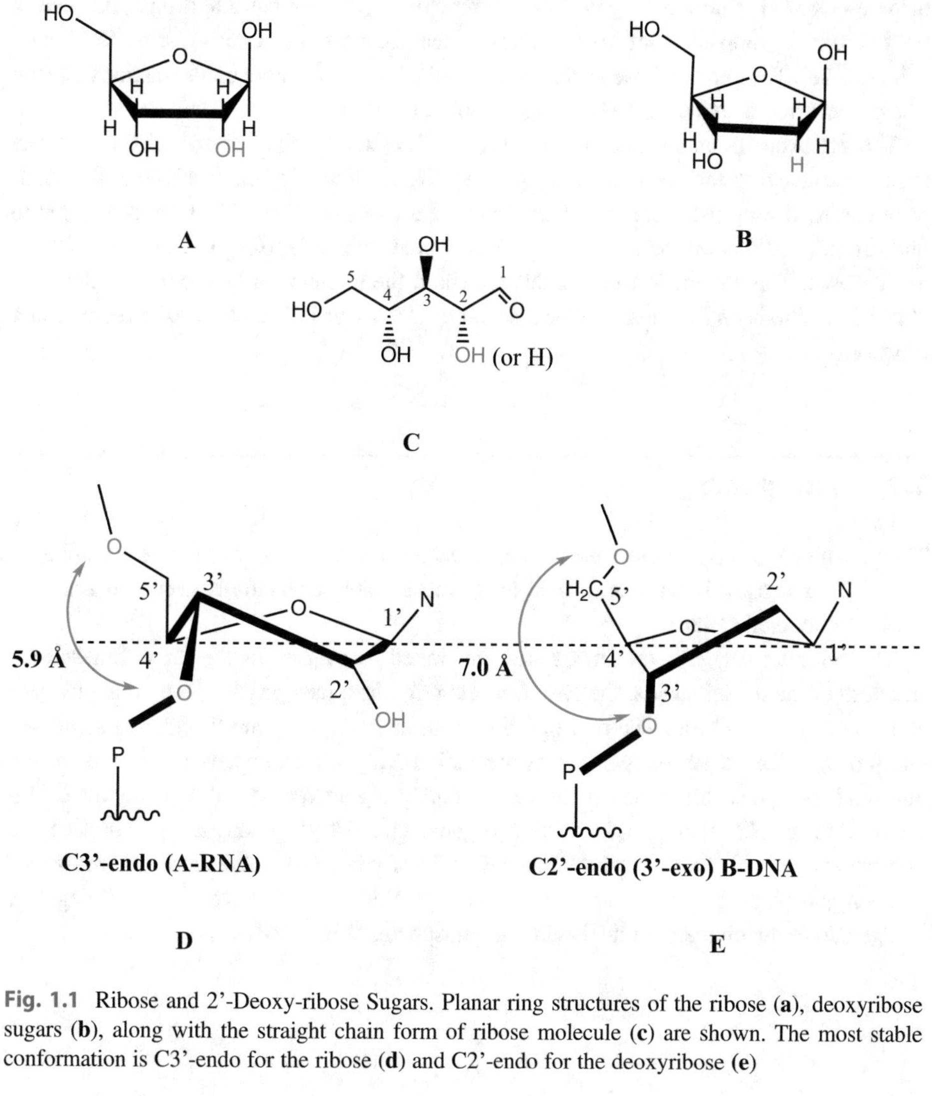

Fig. 1.1 Ribose and 2'-Deoxy-ribose Sugars. Planar ring structures of the ribose (**a**), deoxyribose sugars (**b**), along with the straight chain form of ribose molecule (**c**) are shown. The most stable conformation is C3'-endo for the ribose (**d**) and C2'-endo for the deoxyribose (**e**)

In a planar ring structure, the bond electrons on each carbon eclipse each other, creating repulsion between neighboring electrons. The sugar rings are therefore prefer puckered (nonplanar) conformations, which allows the electron clouds to become staggered, relieving the repulsion strain.

The molecules spend more time in those conformations that are energetically favorable. The ribose sugar of RNA and deoxyribose sugar of DNA have four carbons in the ring with one oxygen occupying the ring position. Carbon $5'$ is outside the ring and the oxygen of 4'O is now part of the ring structure. Various ring conformations are possible as the molecular structures are dynamic. The most stable conformations of ribose involve displacing at least one atom out of the plane defined by C4'-O-C1' atoms (Fig. 1.1d, e) [1].

The most stable conformation for ribose sugar is $3'$-endo, where C3' lies above the plane made by C4'-O-C1' atoms (Figure 1.1d). When the $2'$-position has a hydrogen (instead of the hydroxyl) group attached to it (deoxy), then the preferred conformation is $2'$-endo (Figure 1.1e). The deoxy ribose sugar can theoretically sample more conformations that the ribose sugar and is expected to be more flexible than ribose.

The sugar positions are numbered based on the straight chain form, with the lowest number assigned to the aldehyde or keto group. In addition, a prime is added to the sugar positions to distinguish these from the nucleobase atoms. Thus, $2'$, $3'$, and $5'$ refer to the carbon positions on the sugar and 2, 3, 4 would refer to corresponding atoms on the nucleobases. This distinction makes talking about the nucleic acid structures easier. The $2'$ and $3'$ positions on the sugar will be discussed often as we talk about RNA structures and functions.

1.3   Phosphate

The phosphate groups connect the sugars to each other via two phosphodiester linkages (Fig. 1.2). The ester linkage is formed by losing a water molecule between an acid and alcohol functional groups.

The bridging oxygens on phosphates are linked to sugars. In the three-dimensional structure of the nucleic acids, the two non-bridging phosphate oxygens are nonequivalent and hence, are pro-chiral. The non-bridging phosphate oxygens are designated as pro-R_p and pro-S_p. When these oxygen atoms are individually substituted by another atom, the phosphate becomes chiral; the substituted sites are designated as R_p or S_p (using the Cahn-Ingold-Prelog rules that you learned in organic chemistry), where the p indicates the phosphate atom [2]. These oxygens behave differently in RNA interactions and reactions.

Each phosphate group has one negative charge. Nucleic acids have an overall negative charge due to the charges on the backbone phosphates (Fig. 1.2).

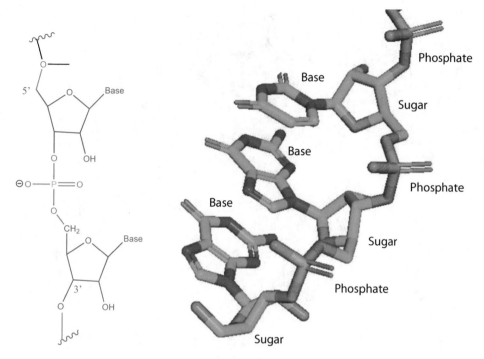

Fig. 1.2 Phosphodiester Linkage. The 5′- and 3′-hydroxyl groups of sugars form ester linkages with the phosphate groups to make the polymer. The crystal structure of RNA (1.5 Å resolution) is shown. The sugars and phosphates form the phosphodiester linkages which are considered the backbone of nucleic acids. Figure made using PDB file 2G91 in PyMol

1.4   Nucleobase, Nucleoside, and Nucleotide

The nucleobases, or bases, contain nitrogen in the heterocyclic aromatic ring structures and are basic. The bases are not directly linked to each other but via the sugar connected to phosphate. A base is attached at $1'$ position on the sugar through a C1'-N linkage (Fig. 1.2).

There are five nucleobases, adenosine (A), guanosine (G), thymine (T), cytosine (C), and uracil (U). Adenine and guanine are purines (R) and have two aromatic rings, each with two nitrogen atoms. The other three bases, uracil, cytosine, and thymine are pyrimidines (Y) and have a single aromatic ring with two nitrogen atoms. The numbering of purines and pyrimidines is based on giving the ring nitrogen atoms the lowest numbers. Learn to draw the bases; know the exocyclic ligands on each base and their positions (Fig. 1.3).

Nucleobases bind to sugars through the nitrogen at position 9 in purines, and position 1 in pyrimidines. Often RNA and DNA are distinguished solely through the difference in their use of nucleobase U or T. This is an important difference but not the most significant difference between the two nucleic acids when it comes to their structures and stability.

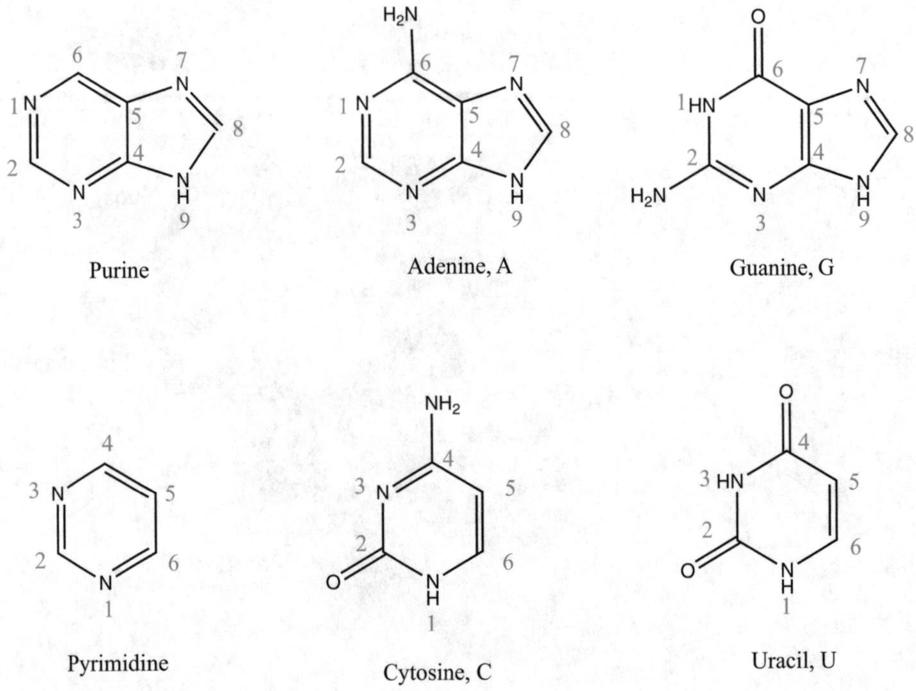

Fig. 1.3 Nucleobases in RNA. The purine (R) bases are guanine (G) and adenine (A). The pyrimidine (Y) bases are cytosine and uracil

When a nucleobase attached to a sugar is a nucleoside. Guanosine, adenosine, cytidine, and uridine are nucleoside corresponding to the four nucleobases. A nucleobase attached to a sugar at the 1′ and a phosphate at the 5′ position on the sugar is a nucleotide (Fig. 1.4).

The tri-phosphorylated nucleotides are needed for the synthesis of DNA and RNA polymers in replication and transcription, respectively, but also play other roles in the cell (Fig. 1.4). Hydrolysis of NTPs is favorable and provides energy for many reactions in the cell. Modified nucleosides such as, dideoxy nucleotides or modifications that alter the 2′ or 3′ positions, serve as drugs against viral polymerases, such as HIV-1 reverse transcriptase, or for chain termination reactions used in DNA sequencing technologies.

In the 1970s a fifth base, 2-amino adenine (Z) (adenine with an additional exocyclic amine at position 2), was identified in a viral DNA and it had replaced all adenines in the genome. Others bases may exist in nature.

1.5   Base Pairing: Canonical and Noncanonical

Canonical Base Pairs The structures of nucleic acids are formed by hydrogen bonding between bases and stacking interactions of the base pairs. With canonical base pairing G pairs with C (G-C) and A pairs with U (A-U). Three hydrogen bonds form between G–C

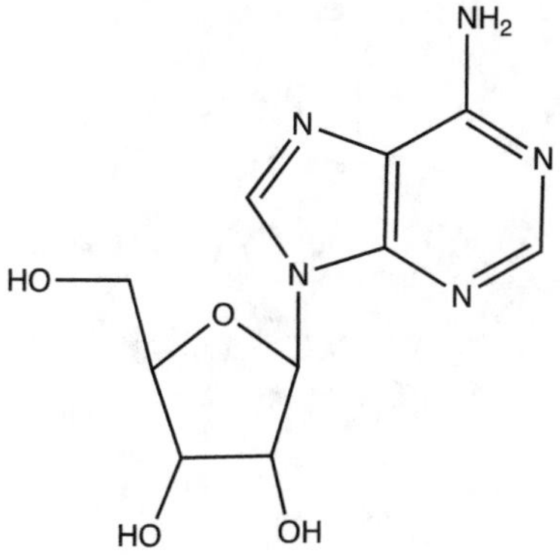

Adenosine 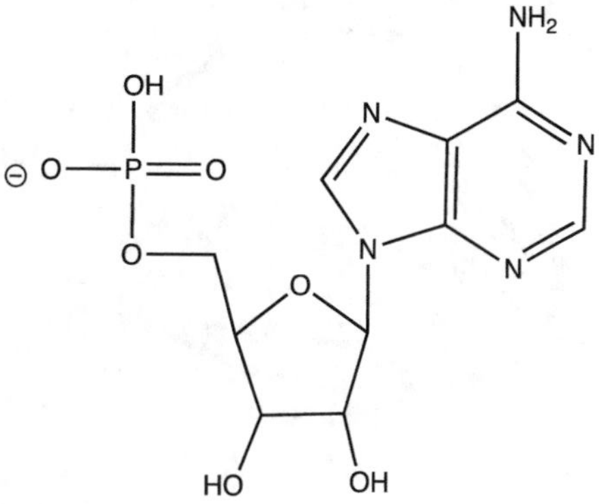   Adenosine monophosphate, AMP

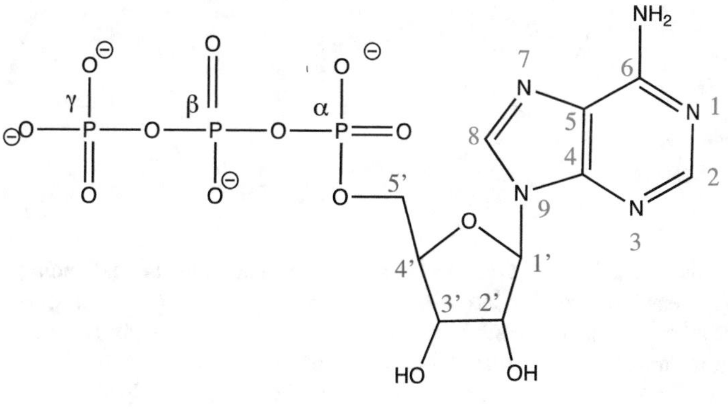

Adenosine triphosphate, ATP

Fig. 1.4 Nucleoside and Nucleotide. A nucleoside is a nucleobase attached to ribose. When adenine is attached to ribose, it is adenosine, which is a nucleoside. Adenosine attached to a phosphorylated sugar is a nucleotide, adenosine monophosphate (AMP) or adenosine triphosphate (ATP)

and two hydrogen bonds between A-U (Fig. 1.5). These two base pairs occupy space in an identical manner and are said to be isosteric; the four combinations (G-C/C-G/A-U/U-A) can substitute for each other to form the regular A-form helical structures in RNA without distortions, more on this in Chap. 2. Note that the 2-amino adenine identified in one virus would base pairs with T (Z-T) using three hydrogen bonds, in a manner similar to G-C.

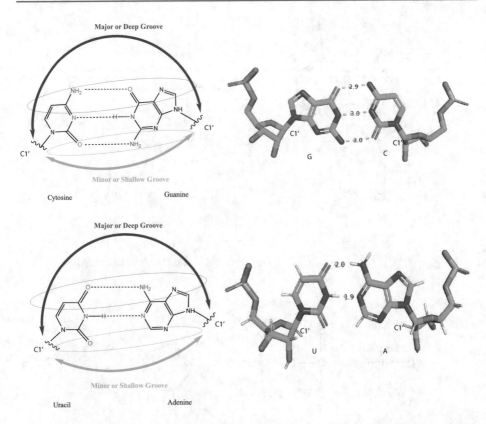

Fig. 1.5 Canonical Base Pairing. Canonical base pairs form upon hydrogen bonding between G-C and A-U. Hydrogens have been added to the U-A base pair but are often not seen in the crystal structures. The atoms that are hydrogen bond donors (blue) and acceptors (red) are color coded in the schematic. The major/ deep groove (dark blue) and minor/shallow groove (light blue) side of the base pair are marked. Shallow groove is more appropriately called the sugar edge

The ligands on the nucleobases are available to hydrogen bond and interact with waters, ions, proteins etc. The top portion of base pairs (as written on the page) are facing out into the region that forms the deep groove in RNA. The bottom portion of base pairs (sugar edge side) are in the shallow groove (more on grooves in Chap. 2). Since the attachment to the sugar isn't symmetrical, there are two distances to measure between C1' to C1' of the two sugars; the longer distance (darker blue) is the deep groove side (Fig. 1.5) and the shorter distance (lighter blue) is the shallow groove side [1]. Draw and cut out G-C and A-U base pairs. Measure the distances between the two C1' atoms as shown in Fig 1.5. (Save the base pairs for building helices in the next chapter.)

Noncanonical Base Pairs The non-canonical base pairs describe all association of bases that are different from those between A-U and G-C base pairs [3–10]. The canonical base pairs occur less than 50% of the time in RNA (Chap. 2, Architecture of RNA). In RNA, many different non-canonical interactions are possible between bases [3–9]. For example, adenine can hydrogen bond to guanine (A•G) or adenine (A•A) (Figure 1.6). The non-canonical base pairs are denoted by a dot (•) between bases.

Noncanonical pairs were first identified in the crystal structure of transfer RNA (tRNA) and were predicted by comparative sequence analysis [5–7]. Francis Crick proposed the Wobble Theory in 1966 to account for the imbalance between the number of codons and number of tRNA [3].

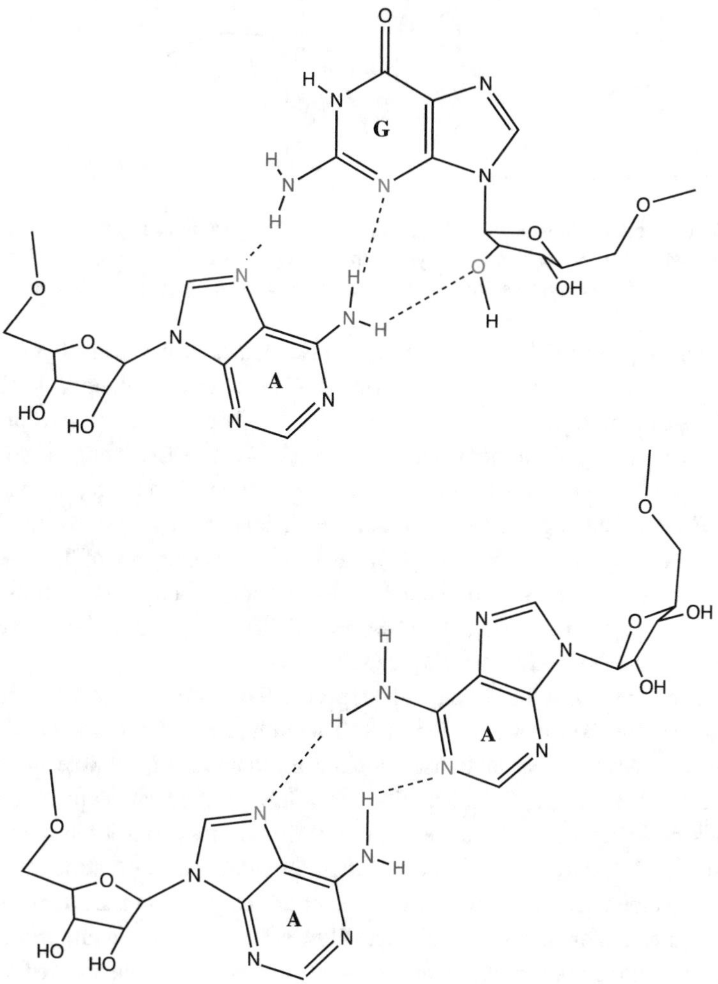

Fig. 1.6 A•G and A•A Base Pairs. The A•G and A•A noncanonical base pairs occur frequently in RNA. Three hydrogen bonds are possible between a A•G base pair with an additional interaction between N1 and the phosphate of the prior residue (not shown). A•A base pairs can form two hydrogen bonds [4]

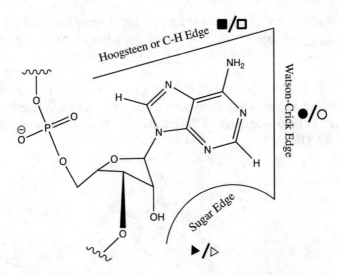

Fig. 1.7 Edges of interactions between base pairs. The filled in symbols represent the cis configuration and open symbols represent trans. In hydrogen bonding, when a Watson–Crick edge is used it is denoted by a circle, the Hoogsteen edge by a square, and the sugar edge by a triangle

A nucleobase presents three distinct edges with which it can form hydrogen bonds [8]. These are designated as: the Watson–Crick (WC, W or canonical) edge, the Hoogsteen edge (HE, H), and the sugar edge (SE, S) (Fig. 1.7). The Watson–Crick edge comprises the positions 2 and 4 in a pyrimidine, whereas a purine's Watson–Crick edge is composed of positions 2, 1 and 6. The Hoogsteen edge describes positions 7 and 8 in purines, and 4 and 5 in pyrimidines. The sugar edge is comprised of interactions with positions near the glycosidic bond. Any base can interact with any other base using any of the three edges to form base pairs. There are a total of 6 combinations of edge-to-edge interactions: WC–WC (WW); WC–Hoogsteen (WH); WC–Sugar (WS); Hoogsteen–Hoogsteen (HH); Hoogsteen–Sugar (HS); and Sugar–Sugar (SS).

The RNA orientation is described with respect to the *glycosidic bonds*, as either *cis* or *trans* as shown by the arrows in Fig. 1.8. Therefore, any of the above edge-to-edge interactions can occur in cis or trans, doubling the number of possible interactions to 12 (Table 1.1). Furthermore, the orientation of the local strands can be parallel or antiparallel. In most cases bases are attached to sugar in anti-orientation and not in syn due to steric constraints (Fig. 1.9). Occasional syn orientation does occur in RNA structures.

In current nomenclature, canonical base pairs are cis-WW and are represented by a closed filled circle. Wobble pairs were originally G•U or those containing inosine (exocyclic amine of adenine is replaced by an oxo—i.e., deamination) and showed a geometric shift between noncanonical base pairs but still mostly formed cis WW base pairs. In current nomenclature, noncanonical base pairs are donated with a dot (•) in the middle or with

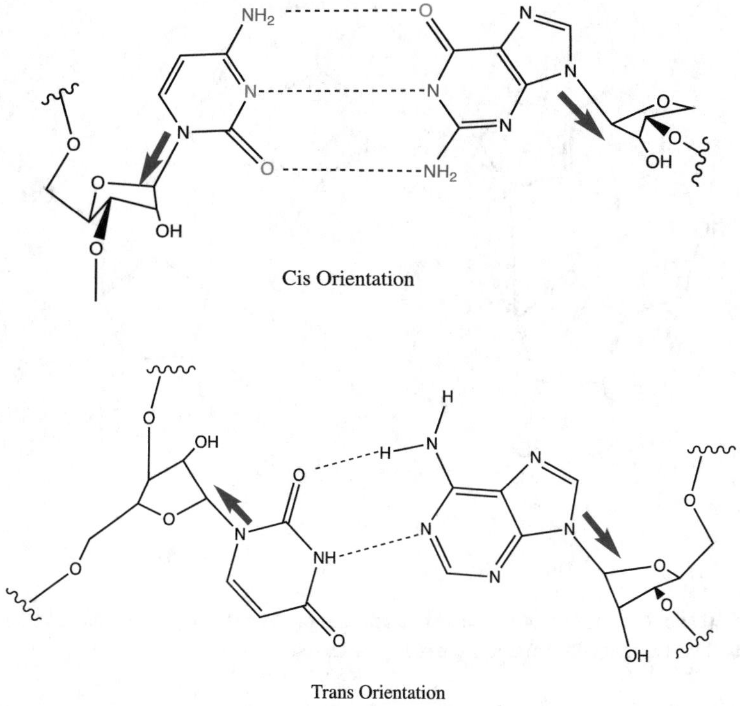

Cis Orientation

Trans Orientation

Fig. 1.8 Cis and trans orientation of glycosidic bonds. The N-C1' linkage is oriented in cis or in trans relative to the hydrogen bonds

Table 1.1 Edge-to-edge notation

Cis-WC/WC	
Trans WC/WC	
Cis-WC/Hoogsteen	
Trans WC/Hoogsteen	
Cis WC/sugar edge	
Trans WC/sugar edge	
Cis Hoogsteen/Hoogsteen	
Trans Hoogsteen/Hoogsteen	
Cis Hoogsteen/sugar edge	
Trans Hoogsteen/sugar edge	
Cis sugar edge /sugar edge	
Trans sugar edge/sugar edge	

The notation used for different types of noncanonical base pairs in Edge-to-Edge format allows for an easy identification of base pairs

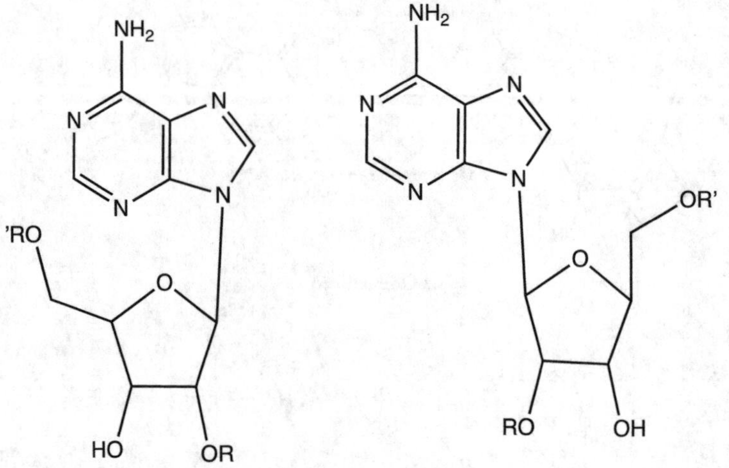

Fig. 1.9 Anti and Syn conformations. The nucleobases prefer anti conformations to better distribute the electron density. Syn conformation are rarely observed

appropriate nomenclature that identifies the type of interactions as proposed by Leontis and Westhof (Table 1.1) [8]. We will use this nomenclature where appropriate.

At least 29 different noncanonical base pair interactions have been proposed and many (~20) of these have been observed [10]. Some commonly seen noncanonical base pairs are: Adenine•Cytosine; Adenine•Guanine; Adenine•Adenine; Uracil•Guanine; Uracil•Cytosine; Uracil•Uracil. Noncanonical base pairings do not all occur at the same frequency within RNA molecules. The most common noncanonical base pairs are the G•U wobble (cis WW) and G•A. Analysis of the G•A base pair transversion (A replacing C) shows that adenine is in the syn orientation with respect to the sugar moiety whereas guanine adopts the trans orientation.

1.6   Base Protonation

Protonation, and thus pKa, of nucleotides are an important consideration for the formation of RNA structures and for the nucleobases acting as general acid or base [11]. Structural data on RNA have shown the presence of protonated nucleobases in RNA. Adenine protonates on the N1 atom, whereas cytosine protonates on N3. Their pKas are 3.8 and 4.3, respectively, indicating that under physiological conditions, protonation is not expected. The local RNA structures provide environments that alter the pKa as is often

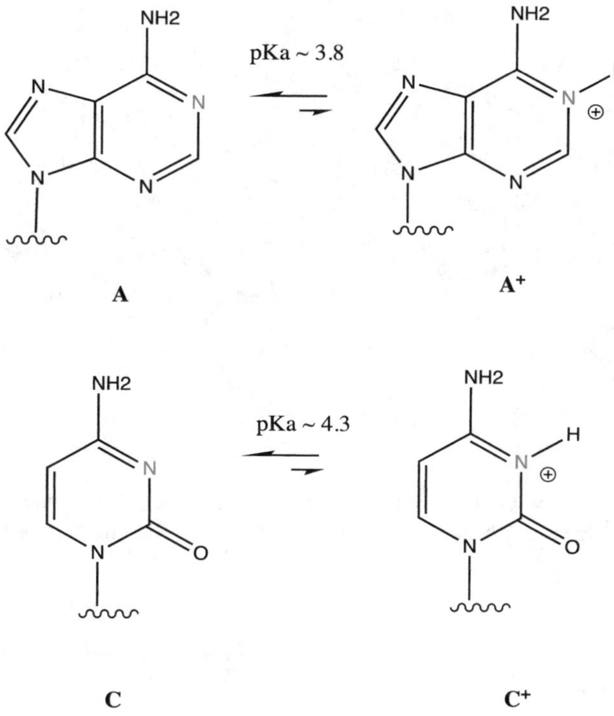

Fig. 1.10 Protonation of adenine and cytosine. In RNA structures, A can be protonated to A+ and C to C+ raising their pKa values to a near physiological pH of 7

the case for amino acids in protein structures. Nucleotides with elevated pKas have been implicated in RNA catalysis (Chap. 3, small catalytic RNA) [11, 12]. Additionally, heavily electronegative areas within the RNA molecule can bring about pKa shifts, indicating the importance of base pair protonation in RNA structural stability (Fig. 1.10).

1.7   Bifurcated and Water-Mediated Base Pairing

Bifurcated hydrogen bonds can form in between bases. Bifurcated systems exhibit three-centered hydrogen bonds—where two H atoms point to a single acceptor [8, 13]. These bifurcated pairs mediate edge-to-edge interactions. For example, the noncanonical G•U base pair is an intermediate *cis*-WC/WC interaction with bifurcated and water-mediated hydrogen bonds.

A "water-inserted" hydrogen bonding is also achieved through the rotation of one base to open a pocket for coordination of a water molecule (Fig. 1.11).

Bifurcated or water-inserted pairs allow for metal ions to coordinate with the RNA backbone. This is expected to occur via the expulsion of the water molecules.

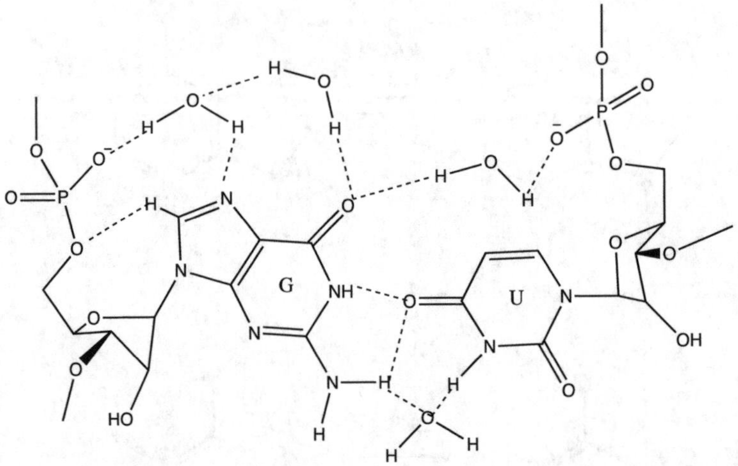

Fig. 1.11 G•U with water-inserted bifurcated hydrogen bonds. The G•U base pair is intermediate cis WC/WC with two hydrogen bonds being formed by the exocyclic groups of guanine and uracil (bifurcated) [13]

1.8 Isostericity

Base pairs are considered isosteric (i.e., occupy the space similarly) when the distances between C1' to C1' atoms and the glycosidic bond angles are similar (Fig. 1.12) [14]. Isostericity allows regularity of helical structures; G-C is isosteric with A-U and substitutions among these do not disrupt the helical structures.

The edge-to-edge nomenclature allows identifying isosteric base pairs. In addition, it allows for pattern identification in three-dimensional structures (Chap. 2, Architecture in RNA).

1.9 Modifications in RNA Bases and Sugars

Naturally Occurring Modification Over 170 modifications have been observed in four nucleobases and in the sugar of RNA (Fig. 1.13) [15]. These modifications are chemical changes to the nucleotide base, or the sugar, that alter the interactions within RNA or with other biomolecules. Modifications are found in all species and within coding and noncoding regions of RNA. The modifications can be reversible or irreversible in nature. Modifying the RNA bases and sugars increases the diversity of functional groups in RNA.

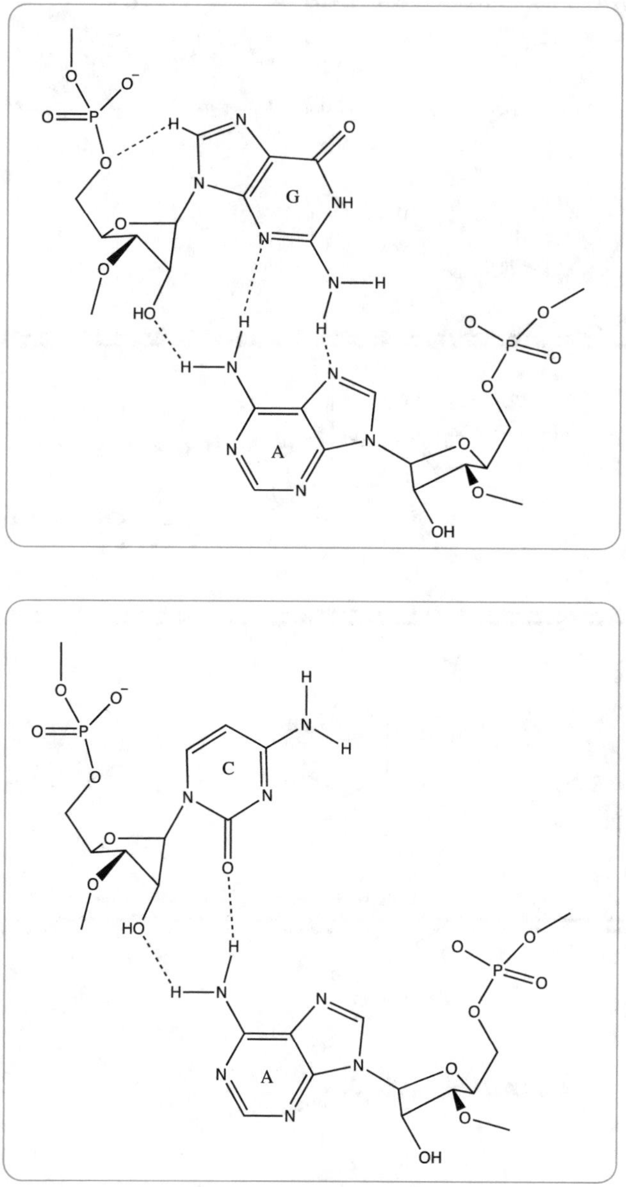

Fig. 1.12 A•G and A•C are isosteric base pairs. In both A•G and A•C adenine is using the Hoogsteen edge and guanine and cytosine are using the sugar edge (A□→G and A□→C). These are trans Hoogsteen/sugar edge base pairs that are isosteric

While some RNA modifications are enzymatically added to nucleotides, other modifications are the result of oxidative stress due to cellular or environmental conditions. The 8-oxo-G modification, for example, is caused by a buildup of reactive oxygen species

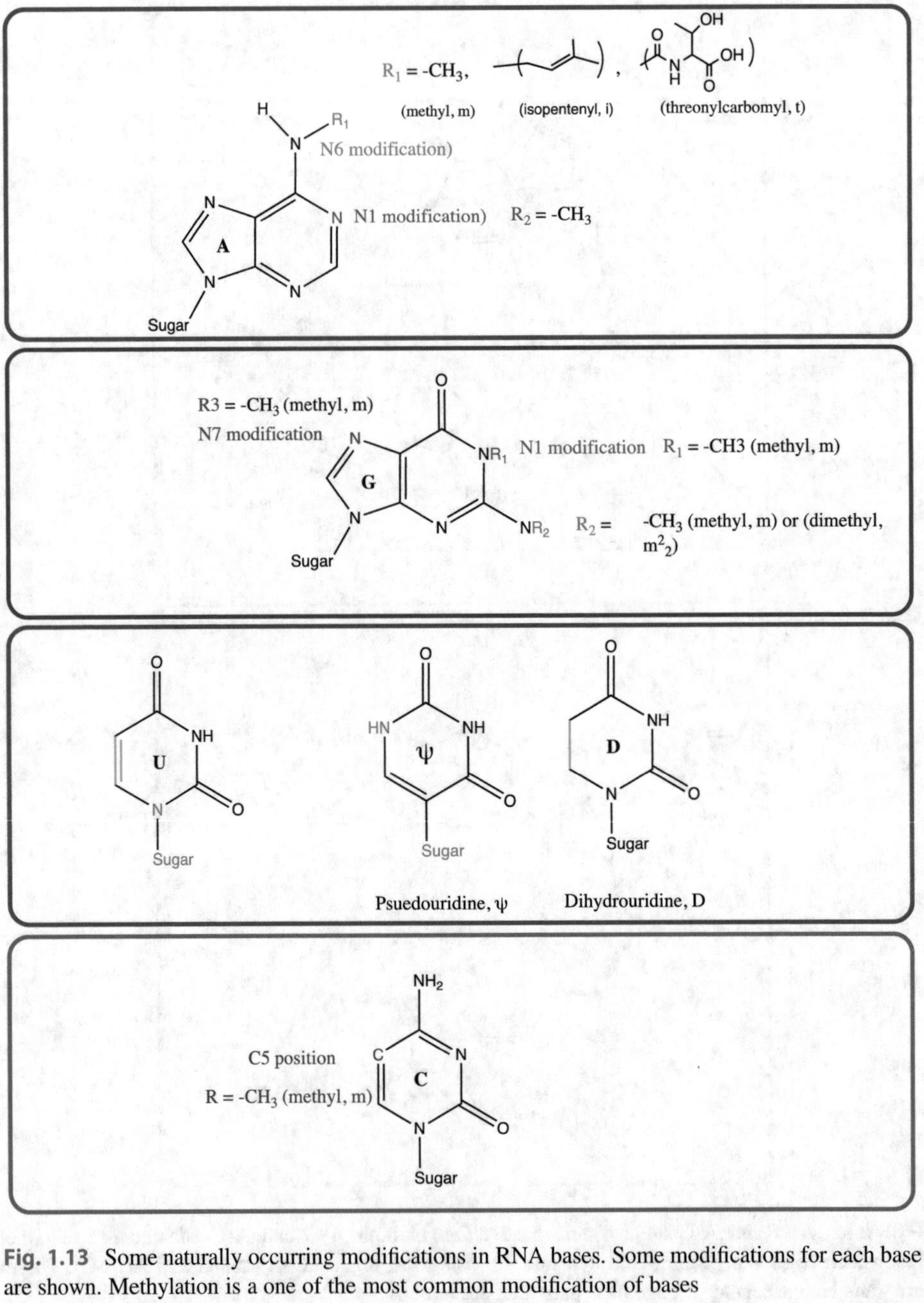

Fig. 1.13 Some naturally occurring modifications in RNA bases. Some modifications for each base are shown. Methylation is a one of the most common modification of bases

and can eventually lead to neurodegenerative diseases. Cells must be able to respond to these RNA modifications rapidly and enable degradation pathways to prevent cellular damage.

Recently RNA have been found on the cell surface bound to sugars (glycosylated RNA) similar to glycosylated proteins and lipids. These RNA likely plays a role in the immune system [16].

Artificial Modification in RNA Along with many naturally occurring modifications, RNA has been artificially modified to alter its chemical and thermodynamic stability. Modifications that stabilize the RNA in the cell make it suitable for in vivo use, for example for RNA-based therapeutics. The 2'-OH is often artificially modified to 2'-deoxy, 2'-O methyl, or 2'-fluoro. Alteration of the 2'-OH to deoxy or 2'-O methyl increases RNA's stability and resistance to degradation by endonucleases. The sugar modifications often alter the conformations of the sugar and the local structures of RNA.

Locked Nucleic Acid (LNA) Sugar modification such as 2'-4' methylene bridge makes RNA resistant to nucleases while maintaining its solubility in aqueous solution (Fig. 1.14) [17]. This modification is called a locked nucleic acid (LNA). The entropic constraint imposed by this modification increases its binding affinity for the complementary strand.

Peptide Nucleic Acid (PNA) The backbone of RNA has also been modified using repeating units of N-(2-aminoethyl) glycine linker to which bases are attached via a methyl carbonyl linker (Fig. 1.14) [18]. Now instead of the naturally occurring sugar-phosphate backbone, a peptide backbone is created. This modification is of particular significance as it alters the charge of the RNA backbone from negative to neutral; thus, there is no electrostatic repulsion upon hybridization, which increases the stability of PNA: RNA strands over RNA:RNA or RNA:DNA hybridization. The PNA, due to its charge

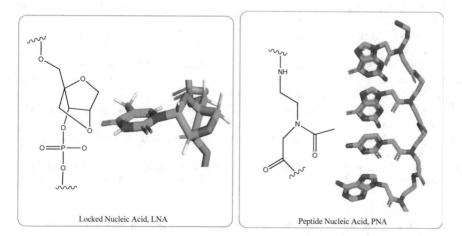

Locked Nucleic Acid, LNA

Peptide Nucleic Acid, PNA

Fig. 1.14 Two artificial modifications to RNA backbone. A 2'-4' methylene bridge locks the sugar to make a locked nucleic acid (LNA). The use of a peptide backbone makes the RNA neutral instead of charged (PNA)

neutrality, also hybridizes in a salt-independent manner. The PNA backbone is resistant to nucleases, thus providing a good substitute for RNA-based drugs.

Take Home Message
- The bases are connected to each other via a negatively charged sugar-phosphate backbone; sugar conformations play an important role in the structures of nucleic acids.
- Canonical base pairs are isosteric and form in cis WW manner.
- Many noncanonical base pairs exist in RNA and can be classified based on the edge-to-edge interactions.
- Natural and artificial modifications to RNA modulate its structure and stability.

References

1. Saenger W. Principles of nucleic acids structures. New York: Wolfram Springer-Verlag Publishers; 1984.
2. IUPAC-IUB Comm. on Biochem. Nomenclature (CBN). Abbreviations and symbols for nucleic acids, polynucleotides, and their constituents. Biochemistry. 1970;9:4022–7.
3. Crick F. Codon—anticodon pairing: the wobble hypothesis. Mol Biol. 1966;9:548–55.
4. Baeyens K, De Bondt HL, Pardi A, Holbrook SR. A curved RNA helix incorporating an internal loop with G•A and A•A non-Watson Crick base pairing. PNAS. 1996;93:12851–5.
5. Suddath FL, Quigley GJ, McPherson A, et al. Three-dimensional structure of yeast phenylalanine transfer RNA at 3.0 angstroms resolution. Nature. 1974;248:20–4.
6. Hermann T, Westhof E. Non-Watson-Crick base pairs in RNA-protein recognition. Chem Biol. 1999;6:R335–43.
7. Gautheret D, Konings D, Gutell RR. G•U base pairing motifs in ribosomal RNA. RNA. 1995;1: 807–14.
8. Leontis NB, Westhof E. Geometric nomenclature and classification of RNA base pairs. RNA. 2001;7:499–512.
9. Masquida B, Westhof E. On the wobble G•U and related pairs. RNA. 2000;6:9–15.
10. Nagaswamy U. Database of non-canonical base pairs found in known RNA structures. Nucleic Acids Res. 2000;28:375–6. https://doi.org/10.1093/nar/28.1.375.
11. Tang CL, Alexov E, Pyle AM, Honig B. Calculation of pKas in RNA: on the structural origins and functional roles of protonated nucleotides. J Mol Biol. 2007;366:1475–96.
12. Thapiyal P, Bevilacqua PC. Experimental approaches for measuring pKas in RNA and DNA. Methods Enzymol. 2014;549:189–219.
13. Varani G, McClain WH. The G•U wobble base pair: a fundamental building block of RNA structure crucial to RNA function in diverse biological systems. EMBO Rep. 2000;1:18–23.
14. Stombaugh J, Zirbel CL, Westhof E, Leontis NB. Frequency and isostericity of RNA base pairs. Nucleic Acids Res. 2009;37:2294–312.
15. Boo SH, Kim YK. The emerging role of RNA modifications in the regulation of mRNA stability. Exp Mol Med. 2020;52:400–8.
16. Flynn RA, Pedram K, Malaker SA, et al. Small RNAs are modified with N-glycans and displayed on the surface of living cells. Cell. 2021;184:3109–24.

17. Koshkin AA, Nielsen P, Meldgaard M, et al. LNA (locked nucleic acid): an RNA mimic forming exceedingly stable LNA:LNA duplexes. J Am Chem Soc. 1998;120:*13252*–3.
18. Nielsen PE, Egholm M, Berg RH, Buchardt O. Sequence-selective recognition of DNA by strand displacement with a thymine-substituted polyamide. Science. 1991;254:1497–500.

Architecture of RNA

2

Hallie Hintz, Ethan Moore, Darryl McCaskill, and Neena Grover

Contents

Keywords

RNA secondary structures · RNA tertiary structures · RNA bulges · RNA ion interactions · Magnesium in RNA · RNA junctions

---

What You Will Learn

A single strand of RNA folds upon itself to form its many structures. The helices and loop regions formed in the secondary structures interact with each other to form a more compact tertiary structure. The charges of the phosphate backbone are neutralized by the potassium and magnesium ions to allow folding of this negatively

(continued)

H. Hintz · E. Moore · D. McCaskill · N. Grover (✉)
Department of Chemistry and Biochemistry, Colorado College, Colorado Springs, CO, USA
e-mail: e_moore@coloradocollege.edu; c_mccaskill@coloradocollege.edu;
ngrover@ColoradoCollege.edu

© Springer Nature Switzerland AG 2022
N. Grover (ed.), *Fundamentals of RNA Structure and Function*, Learning Materials in Biosciences, https://doi.org/10.1007/978-3-030-90214-8_2

charged polymer. In this chapter, we will learn the vocabulary of secondary and tertiary structural features in RNA and discuss the roles of magnesium ions. RNA structures are modular and are likely to be made of patterns of structures that repeat. We will discuss examples of some RNA motifs, the repeating structural units, that are known.

Learning Objectives

After completing this chapter, the students should be able to:

- Discuss A- and B-form helical structures of nucleic acids.
- Describe the role of noncanonical base pairs in RNA structures.
- Define structural elements that are routinely formed in secondary structures of RNA.
- Begin recognizing patterns in 3D structures of RNA, such as a cross-strand purine stacks and pseudoknots.

2.1   Introduction

The primary structure of a nucleic acid is the sequence of nucleobases written in 5′ to 3′ direction. The sugar-phosphate backbone of the nucleic acid is a constant and therefore not written down unless needed for clarity. For example, GpC represents a G next to a C in a sequence whereas G-C represents a base pair between a specific G and C that are distant in sequence from each other; these are often accompanied by numbers to indicate their position in the sequence, for example, G30-C45.

The folding of an RNA is sequence-dependent. The sequence of RNA dictates its secondary and tertiary structures. The RNA structure is composed of double helices interspersed with loops of various sizes. Most helical segments in RNA are short, about 8–10 nucleotides. Helical stems often contain noncanonical base pairs and are connected to the loop regions. The loop segments may link multiple helical segments. Stacking of neighboring helices (coaxial stacking) is an important element of RNA structural stability. Further interactions between distant regions of the RNA help fold the RNA into a more compact structure.

The secondary structures of RNA are predicted by comparing thermodynamic stability of various structures that can be formed using a given sequence (free-energy minimization) and by comparing the sequence variation between organisms (phylogenetic analysis), among other methods [1–3]. Secondary structure predictions based on the thermodynamics are based on data collected on small RNA constructs, often in 1 M salt [3]. Double-helical regions of RNA are easier to predict as the rules for forming canonical base pairing are well understood. The rules for structures formed in non-helical regions are also being elucidated. When sequences of different organisms are compared to each other, patterns of covariations and nucleotide conservation emerge forming the basis for phylogenetic analysis [4–8]. Phylogenetic analysis recognizes the regions of RNA that interact with each other; it identifies residues that are conserved, and therefore, relevant for a particular structure or function.

In the last two decades, the structures of many small and large RNA have been solved. In addition, the secondary structures of RNA are being mapped in vivo in many different organisms. In this chapter, we will discuss the vocabulary of secondary and tertiary structural elements and introduce the concept of motifs. Throughout the book, we will use the language of RNA structures discussed here.

RNA, like proteins, are primarily structured with some variable regions. RNA, in addition, can adopt multiple similar energy conformations in solution. Different conformations of RNA are likely to play important cellular roles. The final distribution of RNA structures in solutions is determined by the RNA–RNA, RNA–proteins, RNA–ions, and RNA–small molecule interactions, in short, by its environment. The flexibility of RNA to form many different isoenergetic structures determines its cellular functions; some of this structural flexibility will be discussed as we discuss particular structures in coming chapters.

2.2 Secondary Structures in RNA

A single strand of RNA folds upon itself to form its many structures. The stacking, hydrogen bonding, and additional interactions with water and ions all contribute to the formation of RNA structures. We will first discuss the many structural features that form locally in a given sequence, these are the secondary structures.

Helical Structures RNA forms the A-form helical structures which are different from the B-form helical structures in DNA in significant ways [9]. Table 2.1 and Fig. 2.1 show the differences between the A-and B-helical forms that occur due to the differences in sugar pucker and additional hydrogen bonding due to the 2'-OH in RNA. Both A- and B-form helices are right-handed and are stabilized by stacking and hydrogen bonding interactions between bases to form base pairs. Other helical structures also form under different ionic conditions but are not discussed here.

An A-form RNA helix is 23 Å in diameter with 11 base pairs per helical turn; each turn is 28 Å; the bases are at a ~ 19° angle to the helical axis (titled steps on a ladder) with 5.9 Å phosphate-to-phosphate distance due to a more rigid 3'-endo conformation of the sugars. Helical regions in RNA secondary structure are marked by P to denote paired regions. Paired regions include canonical or non-canonical base pairs and are also called stems.

Table 2.1 Helices in RNA and DNA

	A-form helix	B-form helix
Handedness	Right	Right
Base pairs per turn	11	10
Rotation/base pair (twist angle)	33°	36°
Inclination of base pair to the axis	+19°	-6°
Rise/base pair	2.6 Å	3.4 Å
Pitch/turn	28.2 Å	33.2 Å
Glycosyl angle	Anti	Anti
Sugar pucker	C3'-endo	C2'-endo
Diameter	23 Å	20 Å

A RNA single strand folds to form short A-form helical regions interspersed by other structural elements. DNA helix is mostly formed by two complementary strands coming together to form a continuous B-form helix. RNA helix is shorter, wider, and more compact, with base pairs that are titled relative to the central axis

A-form helices have a deep groove that is narrower than the major groove of B-DNA; it also has a shallow groove that is wider than the B-DNA's minor groove. In A-form helices, the richness of the deep groove hydrogen bond donors and acceptors is inaccessible to proteins. The narrow groove provides proteins access to the base ligands from the sugar edge side. In RNA, the interactions with proteins and other biomolecules often occur in regions where the backbone adopts a different conformation, due to noncanonical base pairs or other structural motifs, as discussed below. (In RNA, the grooves are often called major and minor, but note that helical structures and grooves are different in the two forms; it is perhaps more appropriate to use the deep and shallow nomenclature for the grooves in RNA.) RNA helical structures are also more rigid.

The B-DNA helix is 20 Å in diameter and has approximately 10.5 base pairs per helical turn in 34 Å; the bases are nearly perpendicular to the helical axis forming a clearly defined major and minor groove with 7.0 Å phosphate-to-phosphate distance. The B-form helix is less rigid as the sugar in DNA is more flexible.

In DNA, two strands are involved in forming a double-helical structure. A 5'to 3' strand base pairs with a 3'to 5' strand and the two strands are complementary, that is, G base pairs with the C (G-C) and A with T (A-T)—these pairings are called canonical pairing. Often the sequence of only one strand is reported in the 5' to 3' direction as the complementary strand sequence can be inferred by canonical base pairing. DNA's B-form helical structure is famously reported in photo 51 taken by PhD student Raymond Gosling, working under the direction of Dr. Rosalind Franklin [10]. This X-shaped pattern on the X-ray is seen for the helical structures. The photo was shown to Watson and Crick, without Franklin's permission, which led them to modify their model to place the base pairs in the middle and the sugar-phosphate backbone on the outside, opposite of what they were trying before. They did not give Dr. Franklin credit for her work in solving the structure of DNA [10].

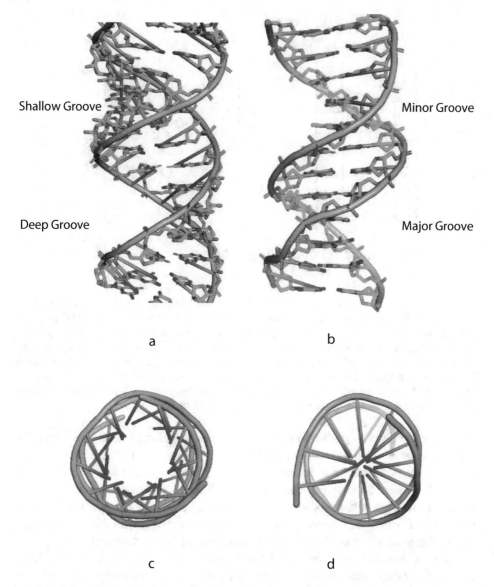

Shallow Groove

Minor Groove

Deep Groove

Major Groove

a

b

c

d

Fig. 2.1 A-form RNA and B-form DNA helices. The RNA A-form helix (**a**) and DNA B-form helix (**b**) are both right-handed helices. The number of base pair per turn, the angle of base pairs relative to the central helical axis, and the grooves are different in RNA and DNA helices. Looking down the helices from the top—A-form in (**c**) and B-form in (**d**)—show the differences in base pair packing in the middle of the helices

Secondary Structures beyond the Helix When RNA folds upon itself to form structure, it forms many helical regions that contain canonical and non-canonical base pairs along with hairpins (also known as stem loops), internal loops, and bulge loops (Fig. 2.2) [11–13]. This

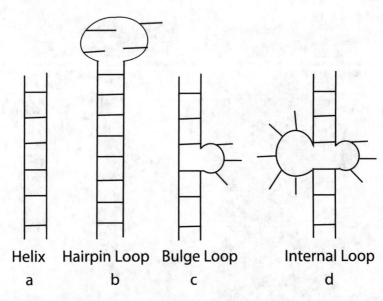

Fig. 2.2 Some common secondary structures in RNA. (a) The double helix is the region of the standard A-form helix. When it contains non-canonical base pairs as well, it is referred to as paired region or stem (P). (b) The hairpin loops (L) are the turning points in the RNA strand that allows for double-helical structures to form a stem. (c) The bulge loops form when one strand is longer than the other. A three-nucleotide bulge loop is shown above. (d) The internal loops form when both strands have noncanonical base pairs in a helix; these loops can be either symmetric (n x n) or asymmetric (n x m) based on the number of bases on each strand. An asymmetric 4 x 3 internal loop is shown above. Both bulge- and internal-loop are also two-way junctions (J)

level of structure can be predicted using the 5′-3′ sequence information alone using phylogenetic comparisons or free-energy minimization [1–3]. The non-helical regions are also structured even when drawn as unstructured "bubbles" in many secondary structures [1–8, 11–13]. Stacking of bases plays a significant role in forming RNA structures. It is important to note that stacking of bases causes exclusion of water between base stacks. The structures of non-helical regions are harder to predict as they are context dependent. Since these are the regions that are involved in interactions with biomolecules, the structures that are recognized may be conserved but the sequence that forms these structures may not be. In addition, RNA in these regions may represent several different conformational states that are then selected based on further interactions. Structural methods, including analysis of deposited crystal and NMR structures, and in-vivo structural probing methods, are increasingly being utilized to determine the rules of structure formation in RNA [14, 15].

The ability to predict secondary structures using RNA stability (free-energy minimization) greatly enhanced our understanding of RNA structures [3]. Stacking between base pairs, stacking of unpaired bases, hydrogen bonding, and other interactions with water and ions all play significant roles in forming the structures.

Binding of metal ions to the negatively charged backbone is essential for the formation of both the local structure and for the folding of the larger RNA [16–18]. RNA helices bind

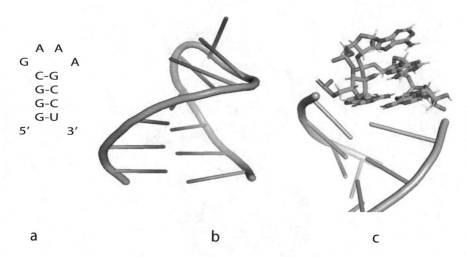

Fig. 2.3 Hairpin Loop. The hairpin loop forms when RNA backbone fully turns. A hairpin loop's secondary structure (**a**) and NMR structure (**b**) are shown. The structure of a GNRA tetraloop shown in cartoon form [19]. (**c**) The interactions within the tetraloop show a G•A base pair (trans sugar/WC) with purine (R) of the loop stacked on top of the base pair; N may be stacked, as seen with adenine, or unstacked depending on the nucleotide occupying this position. The figure was made using PDB file 1ZIF in PyMol

ions in the deep groove. Specific base pairs, such as G•U have exocyclic oxygens positioned in the deep groove and hence are expected to bind ions and water more readily.

Hairpin Loops/Stem Loops (SL) Hairpin loops or stem loops (SL) form when the single stranded RNA folds upon itself—using a hairpin turn—to form a short loop (Fig. 2.3). The nucleotides in the hairpin loop form noncanonical interactions and are often structured [11–13, 19, 20]. The loops can be of various sizes. Four nucleotide loops (tetraloop) occur more frequently. Among all tetraloops, UNCG, and GNRA occur more often in RNA (N = any nucleotide; R = Purine) (Fig. 2.4). The four nucleotide loops adopt specific conformations, such as the U-turn or a Z-turn [20]. Different size hairpin loops may adopt similar three-dimensional structures.

The hairpin loops are involved in many long-range interactions important for RNA folding and its stability. Stem loops play an important role in interactions with proteins and influence diverse biological processes, including rates of transcription, transcription termination, and susceptibility to nucleases (Chap. 10, Transcription) [22].

Bulge Loops A bulge loop is an asymmetrical loop where one strand contains additional bases (Figs. 2.2 and 2.3) [21, 23–25]. These loop nucleotides can be bulged out of the helix or stacked within the helix. When the loop nucleotides are stacked within the helix, it causes the RNA to bend. When the extra nucleotides are pushed out of the helix, it allows the overall geometry to remain close to an A-form helix, thus increasing RNA's stability by

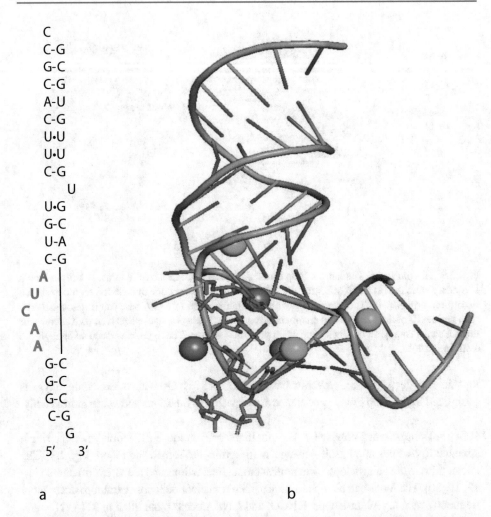

Fig. 2.4 Bulge loops can bend the RNA. (**a**) The secondary structure of internal ribosome entry site (IRES) domain derived from hepatitis C virus. (**b**) The crystal structure of the IRES domain (3 Å resolution) shows RNA is bent by the bulge loop (red nucleotides). It is stabilized by magnesium (green) and manganese (purple) ions. Figure made using PDB file: 2NOK in PyMol [21]

co-axial stacking. The number of nucleotides in the bulge loop and the closing base pairs play a role in the various conformations that RNA takes on. Often these unstacked nucleotides form additional interactions such as, packing into a nearby helical groove or acting as a flap residue in a ligand binding site [23]. The dimensions of both deep and shallow groove are altered by the presence of bulge loops, allowing these to become recognitions sites for protein binding. For example, in HIV-1 TAR and RRE bulge loop serve as binding sites for Tat and Rev proteins, respectively.

Bulge loops serve as hinges and wedges in RNA structure [23, 24]. Metal ions and protein ligands play a role in stabilizing different conformations under different cellular

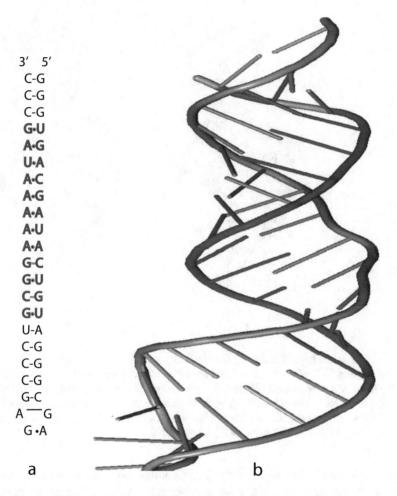

3' 5'
C-G
C-G
C-G
G·U
A·G
U·A
A·C
A·G
A·A
A·U
A·A
G-C
G·U
C-G
G·U
U-A
C-G
C-G
C-G
G-C
A ⎯ G
G ·A

a   b

Fig. 2.5 Symmetrical Internal Loop Distort RNA Backbone. (**a**) The secondary structure of loop E motif has several noncanonical base pairs in a symmetrical internal loop. (**b**) NMR structure of loop E forms a distorted helical structure. The figure was made using PDB file 1MNX using PyMol

conditions [5, 25]. The conformations that perform a particular function are selected by the specific interactions—this is a form of adaptive recognition, a form of induced fit.

Internal Loop An internal loop is a region of noncanonical base pairing that occurs within the helix; these regions can be symmetrical or nonsymmetrical [11–13]. The symmetrical loops often form noncanonical base paired regions and cause distortions in the A-form helical backbone. These distortions to the A-form structure then serve as recognition sites for proteins. Loop E RNA (Fig. 2.5) from the 5S ribosomal RNA has a seven nucleotide

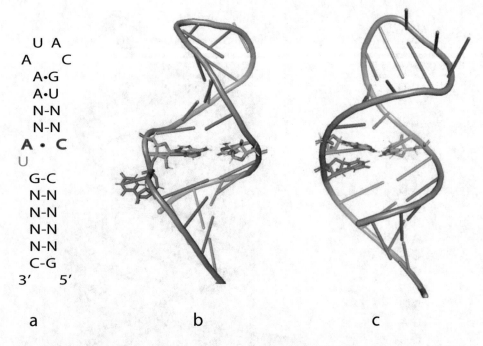

```
    U  A
  A      C
    A•G
    A•U
    N-N
    N-N
  A  •  C
  U
    G-C
    N-N
    N-N
    N-N
    N-N
    C-G
  3'      5'

  a              b              c
```

Fig. 2.6 Asymmetrical Internal Loop. (**a**) The secondary structure of the U6 spliceosomal RNA has a 1 x 2 internal loop. This RNA adopts (at least) two different conformations in two different pHs (5.7 and 7). (**b**) In the NMR structure of U6 at pH 5.7, the A•C pair forms and U (red) is unstacked out of the helix. (**c**) In the NMR structure at pH 7, all the nucleotides of 1 x 2 internal loop are stacked into the helical structure, bending the top stem (note the position of the top loop in **b** versus **c**). Figure made using PDB files 1SYZ and 2KF0 using PyMol

noncanonically base paired region [26]. The breathing rates of canonically paired and noncanonically paired regions might also vary and may allow conformational flexibility.

Asymmetric internal loops form complex and dynamic structures in RNA that have varying stability [26, 27]. A 1 x 2 internal loop occurs within the internal stem loop of a U6 spliceosomal RNA (Fig. 2.6) and is at the heart of RNA catalysis in the spliceosome [28].

The non-helical secondary structural elements change the backbone of RNA in unique ways, and hence, serve as recognition sites for interactions between RNA–RNA, RNA–protein, and RNA–small molecules.

Junctions A junction is a region where multiple helices intersect; the junction region is marked by J and the numbers correspond to the paired regions that link to it [4, 29–31]. Internal loop and bulge loops are two-way junctions. Higher order junctions are seen in many large RNA structures, where three, four, or five helices are seen connected to a junction region.

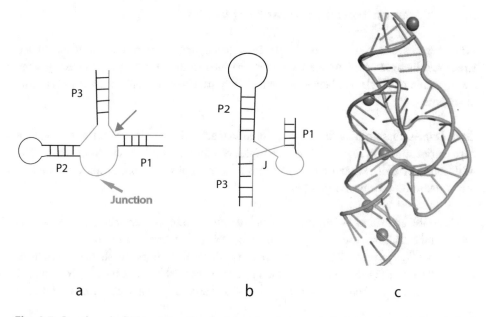

a   b   c

Fig. 2.7 Junctions in RNA. A junction is the region between helical segments. A three-helical junction (blue) is seen in a small RNA enzyme known as the hammerhead ribozyme. The name hammerhead is based on the originally proposed secondary structure (**a**) before stacking of helices around the junction region was known to occur in RNA. The secondary structures are rewritten when the three-dimensional structural information (**b**) becomes available. The secondary structure (**b**) more accurately represents the placement of secondary structural elements in the two-dimensions. The hammerhead ribozyme is an enzyme that cuts its substrate (green); the site of cleavage is shown by the red arrow. The junction region of the hammerhead ribozyme contains some of the most conserved residues and binds the catalytically important metal ion. The three-dimensional structure of hammerhead ribozyme (**c**) shows that the helix 2 and 3 (P2 and P3) stack on top of each other and helix 1 (P1) is stacked upward toward helix 3 with the cleavage site (red arrow) residing in the junction region

Coaxial stacking of helices around the junction makes longer helical segments in RNA possible. Positioning of helices around a junction is constrained by topology of the junction—order (two-way, three-way, etc.), length of connecting strands in the junction, and steric constraints of the interactions. The rules for coaxial stacking and various junction topologies have been proposed and are being classified into different families [4, 29]. For example, in three-way junctions, the length of the linkers between the paired regions determines the position and interactions that occur in the linker [29]. Hammerhead ribozyme (Fig. 2.7) has a three-way junction whereas tRNA has a four-way junction.

2.3 Common Tertiary Interactions in RNA

When regions of RNA that are distant in sequence space interact to form the folded structures, these interactions are called tertiary interactions. RNA is a particularly large and dynamic molecule and hence, a given tertiary structure may only exist under specific conditions.

Base Triples Base triples form when a base interacts with an already existing canonical base pair as shown in Fig. 2.8. The base triples can form in local structures or when RNA folds into its tertiary structure [32]. A repository of all observed base triples is found at: http://rna.bgsu.edu/triples/triples.php

A-Minor Interactions Adenines are a unique nucleobase in their lack of an exocyclic ligand in the shallow groove (sugar edge) side—this is sometimes referred to as adenine's smooth shallow groove. Thus, they can insert themselves into the shallow groove of another base pair [31, 33, 34]. When the unpaired adenosines utilize its O2', N3, and N1 to interact with the shallow (minor) groove of a canonically paired RNA helix, forming

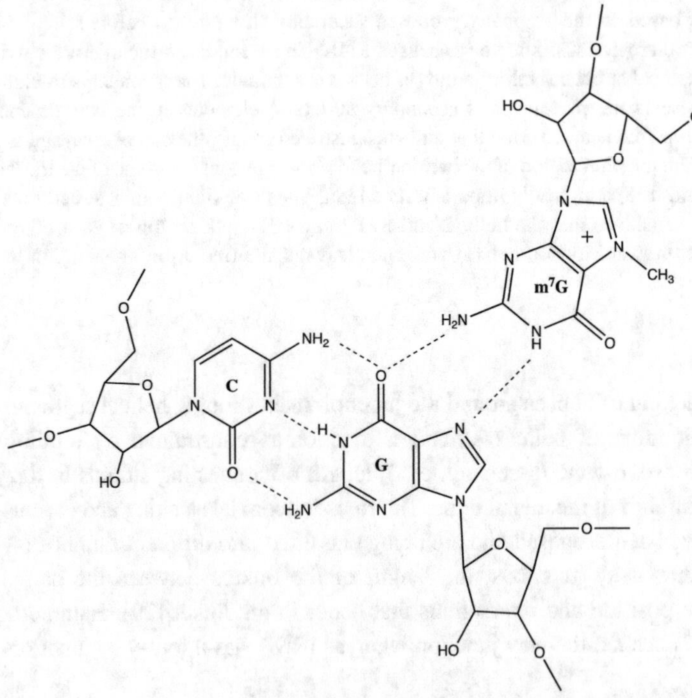

Fig. 2.8 Base triples. A C-G canonical base pair hydrogen bonds with m^7G (N7methyl guanine) to form a base triple. The Hoogsteen edge of G is interacting with the Watson–Crick edge of m^7G

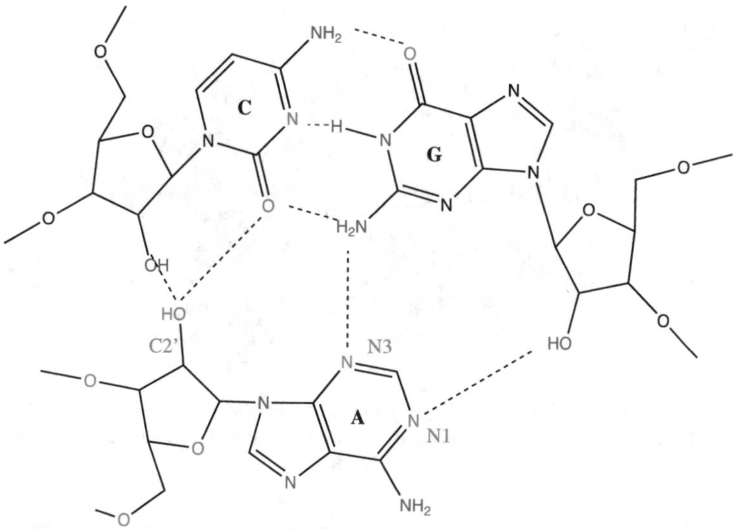

Fig. 2.9 A-minor interactions. Adenine from a distant site is using its N1, N3, C2'-OH to interact with the sugar edge of G-C base pair. The number of hydrogen bonds between adenine and the base pair can vary depending on the position of the adenine relative to the base pair [33]

hydrogen bonds to one or both bases and phosphates, this is called an A-minor interaction (Fig. 2.9). It is a tertiary interaction that allows packing of the RNA structure.

A-minor interactions were first predicted and later found in the structures of hammerhead ribozyme and P4-P6 domain of group I intron. The combination of cis and trans sugar edge/sugar edge (SE/SE) noncanonical base pairs is especially prevalent, especially with adenine. Other nucleotides sometimes participate in similar interactions. The A-minor interactions have been further classified into different types (Type I, II, etc.) based on the position of the inserting base relative to the sugar 2'-hydroxyls of the canonical base pair [31]. Think of this as adenine sliding in the minor groove and providing all (N1, N3 and 2'-OH) or some of these ligands to form hydrogen bonds with the canonical base pairs and their sugar–phosphate backbone in the shallow groove.

Ribose Zippers Ribose zippers are common forms of tertiary structures in which two consecutive 2'-hydroxyls form hydrogen bonds with the 2'-hydroxyls of distant sugars (Fig. 2.10) [35]. These are found in antiparallel chain interactions forming bifurcated hydrogen bonds. Almost all ribose zippers link base paired regions (stems) with loop regions or junctions. The ribose zippers stabilize tertiary interactions in RNA because of the strength of the hydrogen bonds formed by the sugars, and it is often referred to as the "glue" that holds the RNA structures together. These have been classified into many different families.

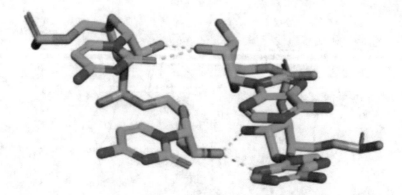

Fig. 2.10 Ribose zippers. The ribose zippers stabilize the overall structure of RNA by forming hydrogen bonds between 2'-hydroxyl groups of two consecutive ribose sugars. Sugars of stem base pairs interact with two sugars 2' from a distant loop region

Kissing Loop and Pseudoknots Base pairs in the hairpin loop often interact with distant regions to form tertiary structures. Kissing loops form when base pairs form between two loops, while pseudoknots form when part of the hairpin loop pairs with a single stranded segment of the RNA (Fig. 2.11 and 2.12) [36, 37]. Pseudoknots are classified into different types. Base pairing and metal ions stabilize these interactions.

The structural features of RNA seen in the crystal structures may be affected by crystal packing and ionic conditions used for experiments. The identity of bound ions in RNA structures should be taken lightly. Even when correctly identified, the sites of metal ions seen in the structure may not be physiologically relevant as RNA is surrounded by many small and large molecules in the cell, many of them are charged. The structural features of RNA are now being explored inside the cell using structural probing methods and genomic technologies. These techniques are already highlighted the differences between in vitro and in vivo RNA structures.

2.4   Examples of RNA Motifs

The structures of RNA are modular. These modular patterns, motifs, that are forming in a larger RNA are being identified and classified [4, 6–8, 11–13, 19–48]. A motif is a three-dimensional structural pattern that results from a particular set of interactions and is found in more than one RNA. Some motifs may be sequence specific, others may have different constraints. Interactions with cofactors, notably water and ions, are an important component of the motifs. The contributions of these additional interactions to the stability of RNA are difficult to quantify.

Motifs can occur within a local secondary structure, such as a hairpin loop motif or these can form upon tertiary folding of secondary structures. Structural elements such as

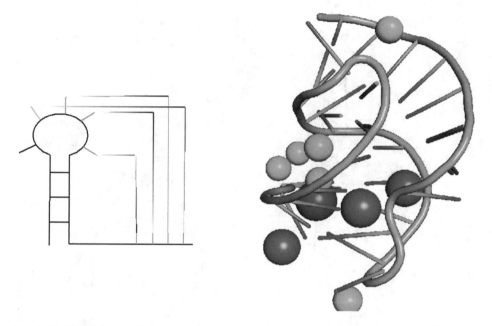

Fig. 2.11 Pseudoknots. A schematic of a H-type pseudoknot is shown along with a 1.25 Å resolution crystal structure of pseudoknot. The pseudoknot shows the loop of a hairpin forms base pairs with a single stranded region of the RNA. Metal ions are seen bound to the RNA (Mg^{2+} green; Mn^{2+} purple) [36]. The high number of ions seen in the structure is likely due to the crystallization conditions. The figure was made with PDB file IL2X in PyMol

particular noncanonical base pairs, particular stacking interactions, and sugar conformations, are components (a subset) of a motif. A key feature of a motif is its ability to form its structure(s) independent of the larger RNA in which it exists. The pattern is often seen in multiple different RNA. A motif is thus a structural property that may have associated sequence requirements. The definition of motif is evolving as we learn more about RNA structures and their classifications.

RNA motifs are sites for interactions between RNA–RNA, RNA–protein, RNA–small molecules and include water and metal ions that interact with it. If various motifs could be identified, then we could "build" the larger RNA structures. This would allow us to predict structures, and associated function, using the sequence data alone. It is theoretically possible to take an RNA virus such as the coronavirus, sequence its genome and predict its various structural and functional components. This would then allow us to devise ways to inhibit the RNA and avert a pandemic. The groundwork for this scenario is being laid at this very moment.

A few examples of motifs are presented below to illustrate the concept of a motif. There are new motifs being discovered regularly due to a rapid growth in three-dimensional structures databases. Recently, machine learning has allowed an AI-based program called

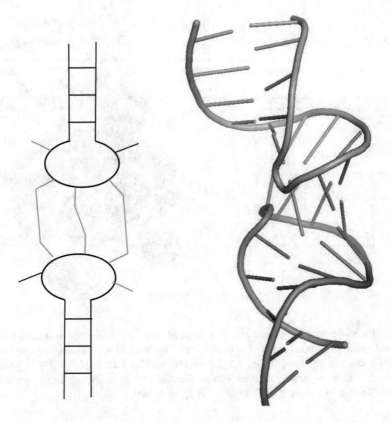

Fig. 2.12 The kissing loop. The kissing loops are formed when the loop nucleotides of two hairpins form base pairs. A NMR structure of TAR-RNA hairpin from HIV-2 bound to a complementary hairpin is shown [37]. The figure was made using PDB file 1KIS in PyMol

AlphaFold to predict, with decent certainty, the tertiary structures of all human proteins. RNA structural predictions are not far behind. It makes it a good time to learn about RNA!

Loop E Motif Bacterial loop E motif was first identified in 5S ribosomal RNA from E. coli [26, 39]. It is a symmetrical internal loop with seven consecutive noncanonical base pairs that only form its structure in the presence of magnesium ions (Fig. 2.13a). It has twofold rotational symmetry, with the axis passing through the fourth base pair; the two 3-base pair sub-motifs are separated by a water-inserted cis W/W A•G base pair.

Both sub-motifs have a G-C base pair followed by cis H/S G•A (also, called a sheared base pair), followed by cis H/W A•U base pair (also called a reverse Hoogsteen base pair). An adenine from a G•A stacks on top of adenine from A•U of the opposite strand, forming a cross-strand purine stack, which alters the shape of the deep groove (~2.1 Å narrower) and shallow groove (~2.2 Å wider). A severe kink in the backbone causes the 2'-OH of G

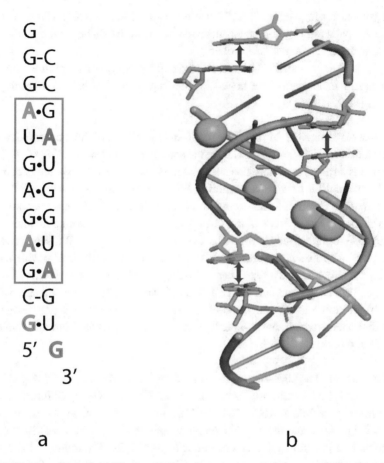

G
G-C
G-C
A•G
U-A
G•U
A•G
G•G
A•U
G•A
C-G
G•U
5′ G
    3′

a   b

Fig. 2.13 The loop E motif. (a) The loop E secondary structure shows seven noncanonical base pairs (boxed). (b) The crystal structure of loop E motif (1.5 Å resolution) shows three cross-strand purine stacks are indicated by dark blue double arrows; purines on one strand are colored blue and the other strand green; magnesium ion (green spheres) bound to the RNA are shown [26]. The figure was made using PDB File 354d in PyMol

of G•A to form a hydrogen bond with exocyclic amino group (N6). In addition, a water molecule interacts with the amino group of the G of G•U base pair and phosphate group of the opposite strand to stabilize this structure.

The non-canonical base pairs between the cross-strand stacked adenines in the two sub-motifs form a single hydrogen bond between bases or a bifurcated hydrogen bond between carbonyl oxygen of one base and imino and exocyclic amino of the partner base; these are also additionally stabilized by bridging water molecules. This might compensate for the stacking interactions missing for the G•A base pair on the 3′ side of this base pair. Unstacking of this G allows space in the deep groove to accommodate a magnesium ion. Five magnesium ions interact with the loop E motif either in hexahydrated, $[Mg(H_2O)_6]^{2+}$, or pentahydrated form, $[Mg(H_2O)_5L]^{2+}$, where L = a ligand on RNA. The two G•A and one G•U base pairs are forming water-mediated bonds in the shallow groove, the site for interaction with the ribosomal L25 protein.

Thus, loop E is a motif that is characterized by cross-strand purine stacks, five hexacoordinated magnesium ions bound to the deep (major) groove and ordered water interactions (Fig. 2.13).

Sarcin-Ricin (or Bulged G-Motif) The sarcin-ricin motif is a highly conserved structural feature of ribosomal RNA and is crucial for the activity of the ribosome [38, 40–42]. When the ribosome is targeted by cytotoxins such as sarcin and α-ricin, translation is completely abolished. This motif is crucial for anchoring EF-G on the ribosome during mRNA-tRNA translocation [42]. This motif also occurs in eukaryal 5S rRNA and is easy to confuse with the bacterial loop E motif. It contains a GAGA hairpin loop and asymmetric internal loop, with a bulged G residue. This G forms a base triple with a U•A (trans WC/Hoogsteen)—all written on the same line in the secondary structure (Fig. 2.14a). The U of this base pair stacks with the A below it. The A•G base pair above the base triple is a trans Hoogsteen/ Sugar Edge. This base triple and the base pairs above and below create a sharp S shape (Fig. 2.14b). The three adenines in the G-bulged region (purple) are stacked to form a cross-strand purine stack—which means exactly as it sounds, the purines from different strands stack on each other (Fig. 2.14c).

Kink-Turns Motif The kink-turn (or k-turn) is a repeated structural motif (Fig. 2.15) first identified in the large subunit (50S) of the ribosomal RNA (rRNA), with further examples seen in riboswitch RNA, mRNA, and snoRNA [43]. Kink-turn, as the name implies, introduces a tight kink into the helical axis often upon binding to ions [24, 30, 43]. It is a motif formed by a junction of a three-nucleotide bulge (L1, L2, L3) followed by G•A, A•G tandem repeats that adopt a trans Sugar edge/Hoogsteen (trans S/H) conformation (also known as a sheared conformation). The stem containing G•A base pairs is called the noncanonical stem (NC stem). The other stem is made up of canonical base pairs (C stem, positions −1, −2, −3). The bulge-containing strand is marked b and the non-bulge-containing strand is marked n. At least two or three non-canonical base pairs are needed to form a kinked structure beyond that of a three-nucleotide bulge loop (NC stem, 1, 2, 3). The first G•A pair (1b•1n) is strongly buckled; the second G•A (2n•2b) pair is closer to coplanarity. The shallow groove edges of the G•A pairs are juxtaposed with the shallow groove of the canonical stem and participate in A-minor interactions.

The two or three trans S/H G•A base pairs achieve the flexibility required in the molecule to kink. It allows the third nucleotide in the loop, L3, to stick out. The conserved adenines (positions 1n and 2b) form cross-strand hydrogen bonds with L1 and -1n. The hydrogen bond donated by 2′-hydroxyl of -1n can form a hydrogen bond with N3 or N1 of adenine. The known kink-turns divide into two types depending on whether they hydrogen bond with N1 or N3 of adenines as this results in difference in the rotation of C-helix.

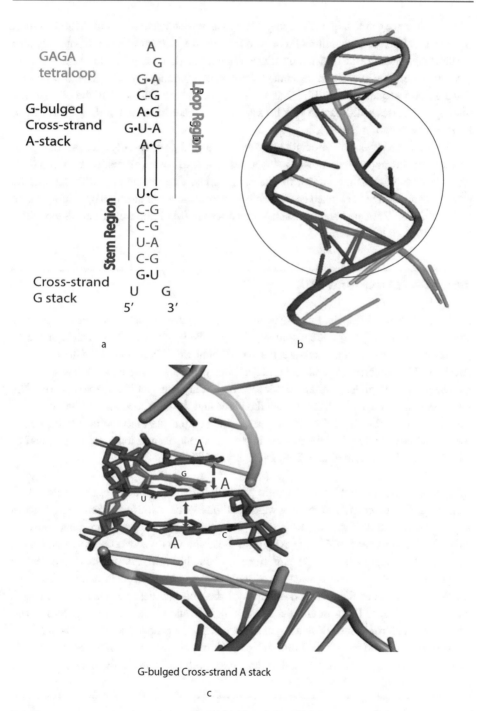

Fig. 2.14 The sarcin-ricin motif. The secondary structure and the crystal structure of sarcin-ricin motif at 1.1 Å resolution are shown. The structure contains a G-bulged region that causes RNA to form an S shape. A base triple formed by G base pairing with a U•A base pair and adenines stacking

In the absence of the larger ribosome, metal ions or proteins that bind to it, the kink-turn is not tightly kinked. Low millimolar or high micromolar concentration of magnesium ions are needed to stabilize the folded kink-turn structure. Position 3b.3n (and to a smaller extent -1b•-1n) determine metal ion-dependent folding. An example of the general kink-turn structure and sequence is in Fig. 2.15a. This three-nucleotide bulge is the site for ion binding and causes a 120-degree kink between stems. A conformation of RNA containing a kink-turn is shown in Fig. 2.15b.

The conserved kink-turn structure is formed by several different sequences and is a great example of the importance of the interactions that lead to the formation of a motif. A majority of kink-turns are known to bind to proteins. The interactions with the specific proteins can stabilize the kinked conformations in the absence of added magnesium ions. In addition, RNA–RNA interactions, such as those seen in SAM riboswitch, can also stabilize the formation of a kink-turn.

2.5   Metal Ions and RNA

RNA molecules fold up to form structures when repulsive forces between negatively charged phosphate groups are countered by positively charged ions. Inside the cell, potassium and magnesium ions are the most abundant. These ions bind to the RNA backbone (diffuse binding) and to particular sites on RNA (site-specific binding) where the charges on the phosphates are closer together. Many site-specific interactions of RNA with metal ions occur with ligands from the bases and the backbone atoms (both on sugar and phosphate) [16–18, 44–53]. Both types of interactions are essential in forming active RNA structures. In addition, metal ions, in particular magnesium ions, are integral part of RNA's catalytic function (Chap. 3, Small catalytic RNA).

Magnesium and Potassium Ions inside the Cell Magnesium and potassium are abundant in the Earth's crust, in the ocean water, and inside the cell. Magnesium salts, such as $MgCl_2$ and $MgSO_4$, are highly soluble at neutral pH. Magnesium ions exist at concentrations around 1 mM in the cytoplasm. Potassium is the most common monovalent cation in the cell and exists at concentrations of about 140 mM. Potassium cations readily interact with RNA structures because of their abundance in the cell. Monovalent ions facilitate charge neutralization and collapse of the phosphodiester backbone; these ions likely form a diffuse layer around the RNA, with ions and water exchanging readily, and with minimal entropic cost. Divalent ions are necessary for proper folding of large RNA. In RNA enzymes (ribozymes), sodium and potassium cations do not seem to be directly involved in catalytic mechanisms. Divalent ions participate in catalytic reaction either

Fig. 2.14 (continued) across strands (marked by double arrows) make this motif a recognition region for ribosomal protein L25 [38]. The figure was made using PDBfile 483d in PyMol

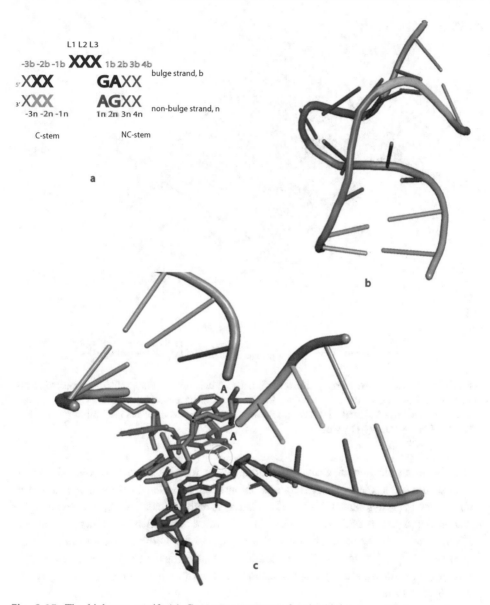

Fig. 2.15 The kink-turn motif. (**a**) Consensus sequence for the kink-turn motif shows a three-nucleotide bulge loop with minimally two G•A base pairs on the 3′ stem. X stand for any nucleotide; the 3b-3n base pair is often a non-Watson-Crick base pair. (**b**) The three-dimensional structure formed by the kink-turn. C) The L1 2′OH hydrogen bond to N1 of A1n is the most critical hydrogen bond (dotted line is circled yellow). Purine stacking in the kink-turn is shown (green and blue adenines are from the non-canonical stem, purple adenine is from the loop). The figure was made using PDB file 7EAG in PyMol

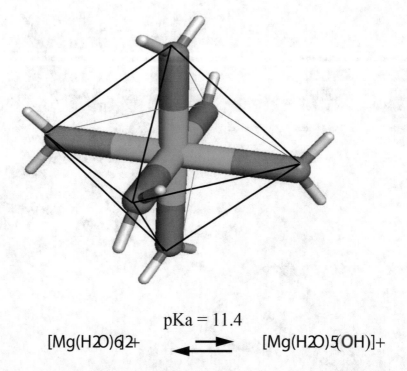

$$\text{pKa} = 11.4$$
$$[Mg(H2O)6]2+ \rightleftharpoons [Mg(H2O)5(OH)]+$$

Fig. 2.16 A hexahydrated magnesium ion. The magnesium ions are often shown as spheres in RNA structures. The magnesium ion is bound to six waters in an octahedral geometry. Hexahydrated magnesium ion has a pKa of 11.4 in free solution. The figure was made using Cambridge Structural Database File 604,009 in PyMol

directly in the reaction mechanism or indirectly in forming the active RNA structures. Magnesium ion's role in RNA stabilization and activation has likely evolved due to its abundance in the cytoplasm. Mg^{2+} has a relatively small ionic radius (0.65 Å) and a higher charge density than Na^+, K^+, or Ca^{2+} and is often referred to as a hard metal ion. Magnesium ions exist in hexahydrated form, $[Mg(H_2O)_6]^{2+}$ with a *slow rate of exchange of water* (10^6 s^{-1}) (Fig. 2.16). For localized ions, magnesium ions are preferred over monovalent cations for charge neutralization, as it is entropically favorable to utilize one Mg^{2+} over two K^+.

Metal Ion–RNA Interactions When fully hydrated monovalent or divalent cations bind with RNA molecules through weak, electrostatic interactions, this is called diffuse binding. Diffuse binding provides charge screening that overcomes electrostatic repulsion between RNA backbone segments and stabilizes the entire molecule. It also accounts for the majority of charge neutralization because of the high mobility and large number of cations available for interactions. Therefore, it is not surprising that the electrostatic potential of

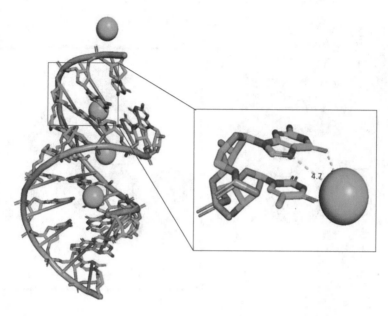

Fig. 2.17 Major groove binding of magnesium ions. Hexahydrated magnesium ions binds in the deep groove of an A-form helix. A 1.4 Å RNA-DNA hybrid structure shows magnesium ions bound in the deep groove and at the mouth of the helix. One of the magnesium binding sites is shown in a greater detail [51]. The figure was made using PDB file 1DNO in PyMol

A-form helix shows significant negative potential in the deep groove as compared to the shallow groove; it is opposite for the B-form DNA helix [51].

Specific interactions with RNA involve direct interactions between magnesium ions and RNA ligands or water-mediated interactions through the first or second shell of waters bound to magnesium ions. Mg^{2+} forms a hexahydrate structure in a strict octahedral geometry. Outer-sphere binding occurs when magnesium cations interact with the RNA backbone and base residues through their first solvation shell. Preferred ligands are anionic phosphate oxygens and some electronegative atoms of bases. Outer-sphere binding is the main Mg^{2+}–RNA interaction seen in structures and is primarily seen in the deep groove of A-form helices (Fig. 2.17).

Inner-sphere interactions occur between RNA and magnesium ions when ligands on RNA replace the bound water on magnesium ion (Fig. 2.18). The local electrostatic environment and the energetics of metal dehydration play a role in direct bonding of RNA and magnesium ions. Partial dehydration of Mg^{2+} is more likely in the non-helical region where neighboring phosphates are in close proximity. Although dehydration of hexahydrated Mg^{2+} is energetically expensive (-1920 kJ/mol), the local environment of RNA plays an active role in making these interactions favorable. The crystal structures of RNA show that specific environments that allow for the formation of multiple direct bonds between Mg^{2+} and RNA. The number of first shell ligands contributed by RNA is seen to

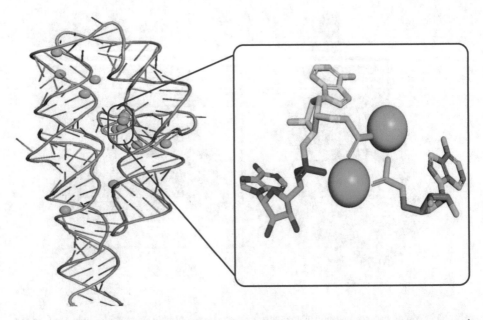

Fig. 2.18 Inner-sphere binding of magnesium ions. The crystal structure the of P4-P6 domain (2.5 Å resolution) of the Group I intron shows several inner-sphere interactions with magnesium. The zoomed in region shows the A-rich bulge with three adenines interacting with magnesium ions; one of the magnesium ions is making three inner-sphere bonds with phosphate oxygens. The figure was made using PDB file 1GID in PyMol

range from one to four in the available crystal structures. Highly chelated Mg^{2+} ions, with two or more first shell ligands, are not abundant but make essential contributions to RNA conformations. These inner-sphere magnesium ions are an essential part of forming the correct RNA structures. Though many ion-binding sites are shown in the deposited structures, the correct ion assignments are nontrivial and are likely to be revised as more careful data analyses are performed [52]. Ions, particular, magnesium ions, play a key role in forming active structures of RNA.

The Role of Metal Ions in RNA Folding Most functional RNA molecules adopt stable, tightly folded globular conformations. A suggested mechanism for RNA folding is the two-stage model. In this model, the RNA folds into its secondary structure first, forming single and double-stranded motifs such as helices, internal loops, and hairpins. This stage of RNA folding is expected to be activated by the presence of monovalent and divalent cations, such as K^+ and Mg^{2+} [16–18, 48]. Next, RNA tertiary structures form, which include many three-dimensional networks of stacked duplexes and intramolecular interactions between the bases and backbone residues. Magnesium ions stabilize tertiary structure by stabilizing secondary structure involved in long-range intramolecular interactions including the interfaces between independently folded segments of RNA

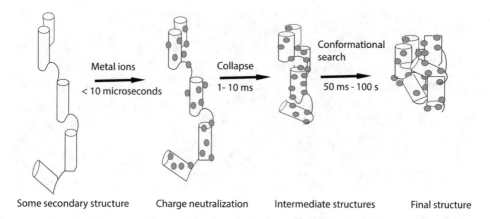

Fig. 2.19 Steps of RNA folding. Charge neutralization by metal ions is an essential step for RNA folding which follows a discreet path to form the active RN structures. Green spheres depict ion binding which neutralizes ~90% of the charge. Figure based on [48]

secondary structures. These networks are essential for the stabilization and specification of an RNA tertiary structure, and allow for functional, native structures to form. Folding into a tertiary structure is strictly regulated by specific metal ion requirements. In order for RNA to collapse into its tertiary structure, magnesium must condense along the secondary structure [48] (Fig. 2.19). Mg^{2+} stabilizes and increases the rigidity of certain RNA conformations that are very unfavorable in the presence of monovalent ions alone, even at high concentrations. Some RNA molecules require greater than 90 percent neutralization of the phosphate charge by metal ions before it folds. Metal ions interact more strongly with folded RNA than unfolded RNA because the negative charge density increases as the phosphodiester backbone folds upon itself.

The Role of Metal Ions in RNA Catalysis RNA can catalyze reactions. The magnesium ions bound to RNA structures are direct or indirect participants in catalysis and are essential components of RNA structures. Catalytic activity in RNA (ribozymes) was first discovered in the 1980s. RNA catalyzes many cellular reactions including protein synthesis and splicing. Early experiments found that some RNA molecules use divalent ions as essential structural and catalytic cofactors. Water molecules directly bound to magnesium complexes participate in catalysis and act as either general acids or general bases by donating protons to the leaving group or accepting protons from the nucleophile. Divalent ion binding also helps to orient and stabilize RNA residues during a nucleophilic substitution reaction. A class of small, self-cleaving ribozymes are catalytically active in the absence of divalent ions but require molar concentrations of monovalent ions to stabilize their structures under in vitro conditions. Most catalytic RNA require metal ions to form their active structures (indirect role). The role of metal ions in the structure and function of RNA is discussed throughout this book (Chap. 3 – Small Catalytic RNA).

Take Home Message
- RNA and DNA form A- and B-form helices that lead to significant differences in grooves, phosphate distances, and interactions with ions, which in turn lead to differences in interactions with biomolecules.
- Nearly half of the RNA structure is non-helical, with many large and small internal and bulge loops, hairpins, and junction regions. Many of these structures require metal ions for charge neutralization.
- RNA structures are modular. Improving our ability to map the structural patterns in RNA will improve RNA structure predictions.

References

1. Woese CR, Magrum LJ, Sigel RB, et al. Secondary structure model for bacterial 16S ribosomal RNA: phylogenetic, enzymatic and chemical evidence. Nucleic Acids Res. 1980;8: 2275–94.
2. Massire C, Jaeger L, Westhof E. Phylogenetic evidence for a new tertiary interaction in bacterial RNase P RNAs. RNA. 1997;3:553–6.
3. Mathews DH, Sabina J, Zuker M, Turner DH. Expanded se-quence dependence of thermody-namic parameters improves pre- diction of RNA secondary structure. J Mol Biol. 1999;288:911–40.
4. Butcher SE, Pyle AM. The molecular interactions that stabilize RNA tertiary structure: RNA motifs, patterns, and networks. Acc Chem Res. 2011;44:1302–11.
5. Mustoe AM, Brooks CL, Al-Hashimi HM. Hierarchy of RNA functional dynamics. Annu Rev Biochem. 2014;83:441–66.
6. Maris C, Dominguez C, Allain FH. The RNA recognition motif, a plastic RNA-binding platform to regulate post-transcriptional gene expression. FEBS J. 2005;272:2118–31.
7. Batey RT, Rambo RP, Doudna JA. Tertiary motifs in RNA structure and folding. Angew Chem Int Ed. 1999;38:2326–43.
8. Strom S, Shiskova E, Hahm Y, Grover N. Thermodynamic examination of 1- to 5-nt purine bulge loops in RNA and DNA constructs. RNA. 2015;21:1313–22.
9. Rich A. The double helix: a tale of two puckers. Nat Struct Biol. 2003;10:247–9.
10. Maddox B, Franklin R. Dark lady of DNA. New York: Harper Collins Publishers; 2003.
11. Hendrix D, Brenner S, Holbrook S. RNA structural motifs: building blocks of a modular biomolecule. Q Rev Biophys. 2005;38:221–43.
12. Moore P. Structural motifs in RNA. Annu Rev Biochem. 1999;68:287–300.
13. Leontis N, Westhof E. Analysis of RNA motifs. Curr Opin Struct Biol. 2003;13:300–8.
14. Boerneke MA, Ehrhardt JE, Weeks KM. Physical and functional analysis of viral RNA genomes by SHAPE. Annu Rev Virol. 2019;29:93–117.
15. Ding Y, Tang Y, Kwok CK, Zhang Y, Bevilacqua PC, Assmann SM. In vivo genome-wide profiling of RNA secondary structure reveals novel regulatory features. Nature. 2014;505:696–700.
16. Draper DE. RNA folding: thermodynamic and molecular descriptions of the roles of ions. Biophys J. 2008;95:5489–95.
17. Pyle AM. Metal ions in the structure and function of RNA. J Biol Inorg Chem. 2002;7:679–90.

18. Record MT, Lohman TM, Haseth PD. Ion effects on ligand nucleic acid interactions. J Mol Biol. 1976;107:145–58.
19. Jucker FM, Pardi A. GNRA tetraloops make a U-turn. RNA. 1995;1:219–22.
20. D'Ascenzo L, Leonarski F, Vicens Q, Auffinger P. Revisiting GNRA and UNCG folds: U-turns versus Z-turns in RNA hairpin loops. RNA. 2017;23:259–69.
21. Dibrov S, Johnston-Cox H, Weng YH, Hermann T. Functional architecture of HCV IRES domain II stabilized by divalent metal ions in the crystal and in solution. Angew Chem Int Ed. 2007;46:226–9.
22. Svoboda P, Di Cara A. Hairpin RNA: a secondary structure of primary importance. Cell Mol Life Sci. 2006;63:901–8.
23. Hermann T, Patel DJ. RNA bulges as architectural and recognition motifs. Structure. 2000;8: R47–54.
24. Schroeder K, Mcphee S, Ouellet J, Lilley D. A structural database for k-turn motifs in RNA. RNA. 2010;16:1463–8.
25. Olejniczak M, Gdaniec Z, Fischer A, Grabarkiewicz T, Bielecki L, Adamiak RW. The bulge region of HIV-1 TAR RNA binds metal ions in solution. Nucleic Acids Res. 2002;30:4241–9.
26. Correll CC, Freeborn B, Moore PB, Steitz TA. Metal, motifs, and recognition in the crystal structure of the 5S rRNA domain. Cell. 1997;91:705–12.
27. O'Connell AA, Hanson JA, McCaskill DC, Moore ET, Lewis DC, Grover N. Thermodynamic examination of pH and magnesium effect on U6 RNA internal loop. RNA. 2019;25:1779–92.
28. McManus CJ, Schwartz ML, Butcher SE, Brow DA. A dynamic bulge in the U6 RNA internal stem-loop functions in spliceosome assembly and activation. RNA. 2007;13:2252–65.
29. Lescoute A, Westhof E. Topology of three-way junctions in folded RNAs. RNA. 2006;12:83–93.
30. Lilley DM. Folding of branched RNA species. Biopolymers. 1998;48:101–12.
31. Lescoute A, Westhof E. The interaction networks of structured RNAs. Nucleic Acids Res. 2006;34:6587–604.
32. Almakarem AS, Petrov AI, Stombaugh J, Zirbel CL, Leontis NB. Comprehensive survey and geometric classification of base triples in RNA structures. Nucleic Acids Res. 2012;40:1407–23.
33. Nissen P, Ippolito JA, Ban N, Moore PB, Steitz TA. RNA tertiary interactions in the large ribosomal subunit: the A-minor motif. Proc Natl Acad Sci. 2001;98:4899–903.
34. Shalybkova AA, Mikhailova DS, Kulakovskiy IV, Fakhranurova LI, Baulin EF. Annotation of the local context of the RNA secondary structure improves the classification and prediction of A-minors. RNA. 2021;27:907–19.
35. Tamura M, Holbrook SR. Sequence and structural conservation in RNA ribose zippers. J Mol Biol. 2002;320:455–74.
36. Egli M, Minasov G, Su L, Rich A. Metal ions and flexibility in a viral RNA pseudoknot at atomic resolution. Proc Natl Acad Sci U S A. 2002;99:4302–7.
37. Chang KY, Tinoco I Jr. The structure of an RNA "kissing" hairpin complex of the HIV TAR hairpin loop and its complement. J Mol Biol. 1997;269:52–66.
38. Correll CC, Wool IG, Munishkin A. The two faces of Esherichia coli 23 S rRNA sarcin/ricin domain: the structure at 1.11 Å resolution. J Mol Biol. 1999;292:275–87.
39. Leontis NB, Westhof E. The 5S rRNA loop E: chemical probing and phylogenetic data versus crystal structure. RNA. 1998;4:1134–53.
40. Szewczak AA, Moore PB, Chang YL, Wool IG. The conformation of the sarcin/ricin loop from 28S ribosomal RNA. Proc Natl Acad Sci U S A. 1993;90:9581–5.
41. Szewczak AA, Moore PB. The sarcin/ricin loop, a modular RNA. J Mol Biol. 1995;247:81–98.
42. Shi X, Khade PK, Sanbonmatsu KY, Joseph S. Functional role of the sarcin-ricin loop of the 23S rRNA in the elongation cycle of protein synthesis. J Mol Biol. 2012;419:125–38.

43. Huang L, Liao X, Li M, Wang J, Peng X, Wilson T, Lilley DMJ. Structure and folding of four putative kink turns identified in structured RNA species in a test of structural prediction rules. Nucleic Acids Res. 2021;49:5916–24.
44. Auffinger P, Grover N, Westhof E. Metal ion binding to RNA. Met Ions Life Sci. 2011;9:1–35.
45. Bowman JC, Lenz TK, Hud NV, Williams LD. Cations in charge: magnesium ions in RNA folding and catalysis. Curr Opin Struct Biol. 2012;22:262–72.
46. Fedor MJ. The role of metal ions in RNA catalysis. Curr Opin Struct Biol. 2003;12(4):555–6.
47. Klein DJ, Moore PB, Steitz TA. The contribution of metal ions to the structural stability of the large ribosomal subunit. RNA. 2004;10:1366–79.
48. Woodson SA. Metal ions and RNA folding: a highly charged topic with a dynamic future. Curr Opin Chem Biol. 2005;9:104–9.
49. Zheng H, Cooper DR, Porebski PJ, Shabalin IV, Handing KB, Minor W. CheckMyMetal: A macromolecular metal-binding validation tool. Acta Crystallogr D Struct Biol. 2017;73(Pt 3): 223–33.
50. Grover N. On using magnesium and potassium ions in RNA experiments. Methods Mol Biol. 2015;1206:157–63.
51. Robinson H, Gao YG, Sanishvili R, Joachimiak A, Wang AH. Hexahydrated magnesium ions bind in the deep major groove and at the outer mouth of A-form nucleic acid duplexes. Nucleic Acids Res. 2000;28(8):1760–6.
52. Leonarski F, D'Ascenzo L, Auffinger P. Nucleobase carbonyl groups are poor Mg2+ inner-sphere binders but excellent monovalent ion binders – a critical PDB survey. RNA. 2019;25:173–92.
53. Chin K, Sharp KA, Honig B, Pyle AM. Calculating the electrostatic properties of RNA provides new insights into molecular interactions and function. Nature. 1999;6:1055–61.

Small Catalytic RNA

3

Jake Heiser and Neena Grover

Contents

Keywords

Catalytic RNA · Ribozymes · RNA cleavage · RNA ligation · Magnesium-RNA reactions

---

What You Will Learn
RNA catalyze biochemical reactions. The endonucleolytic small RNA can cut their own phosphodiester linkages at specific sites. These autocatalytic RNA were first discovered in viroids and later found in a variety of other genomes. In this chapter,

(continued)

J. Heiser · N. Grover (✉)
Department of Chemistry and Biochemistry, Colorado College, Colorado Springs, CO, USA
e-mail: j_heiser@coloradocollege.edu; ngrover@ColoradoCollege.edu

© Springer Nature Switzerland AG 2022
N. Grover (ed.), *Fundamentals of RNA Structure and Function*, Learning Materials in Biosciences, https://doi.org/10.1007/978-3-030-90214-8_3

we will discuss the cleavage and ligation reactions using a few model ribozymes. The roles of magnesium ions, nucleobases, or cofactors as general acid or base will be discussed. Structural patterns seen in these small catalytic RNA are also seen in larger RNA.

Learning Objectives
After completing this chapter, the student should be able to:
- Draw the mechanisms for RNA cleavage and ligation reactions.
- Identify general acids and bases in RNA reactions.
- Explain the direct and indirect role of metal ions, particularly magnesium, in RNA structures and catalysis.

3.1   Catalysis by RNA

For decades, proteins were believed to be the only catalytic macromolecule due to the larger diversity of functional groups available to them. The discovery of catalytic activity in group I intron and RNase P was the start of the era of RNA-based catalysis [1, 2]. The ability of RNA to store genetic information and catalyze reactions supports the "RNA world" hypothesis which proposes RNA as the original genetic molecule for evolution of life as we know it.

Since their discovery in the 1980s, much has been learned about structures and functions of RNA that promote catalysis. RNA-based catalysis is a normal part of cellular process. The catalytic RNA or ribozyme, short for RNA enzymes, are at the heart of splicing and peptidyl transferase reactions (Chap. 4, Spliceosome).

RNA sequences that are significantly shorter than those found in the ribosome and spliceosome are capable of catalysis [3–8]. This class of RNA structures, between ~40 and 200 nucleotides, are grouped into a category of small catalytic RNA. Most are known to perform RNA cleavage and some perform the reverse reaction of ligation. Most use metal ions as cofactors. Some use nucleobases or small metabolites as a general acid or base.

Within the small catalytic RNA class, there is a group of catalytic RNA that are part of the viral genome and are essential in their replication via the rolling circle mechanism (Fig. 3.1). In this mechanism, the circular viral (or viroid) genome is transcribed continuously so that a long single, large, multimeric RNA transcript is formed. This long transcript folds to form small catalytic RNA structures at the junctures that allow cleavage of the multimer into individual monomers. Both the positive and negative strands form the ribozyme structures that facilitate replication using this mechanism.

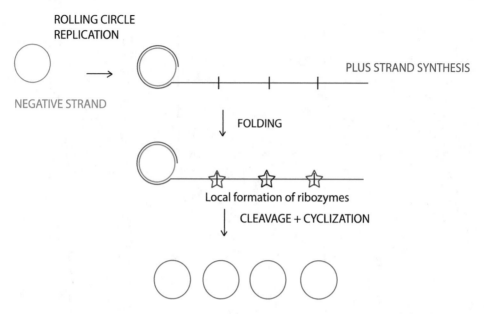

Fig. 3.1 Rolling Circle Replication. The original circular genome in viruses is transcribed into a single long transcript that contains many different copies of the viral genome. Subsequently, the RNA folds to cleave itself into monomers (autocatalysis). Both positive and negative strand are known to form the ribozyme structures

Another class of small catalytic RNA encompasses ribozymes that form in mRNA transcripts. The *glmS* motif in the 5' UTR of an mRNA transcript is a riboswitch RNA that binds a modified sugar [9, 10]. Here the cleavage of the mRNA is a form of gene regulation.

New discoveries via comparative genomic analyses show known and putative ribozymes in mRNA, transposons, and retrotransponson [7].

3.2   Cleavage and Ligation Reactions

All small catalytic RNA discovered thus far catalyze the same two reactions [3–11]. Small catalytic RNA can either cleave the RNA backbone or perform the reverse reaction, the ligation of an RNA backbone via transesterification reaction. Different ribozymes catalyze these two reactions in different proportions, with most favoring the cleavage reaction.

Small catalytic RNA usually act in *cis*, meaning that the strand being cleaved is part of the ribozyme. Because small catalytic RNA cleaves itself, these sequences are often referred to as autolytic or self-catalytic ribozymes. It has been possible (in vitro) to modify *cis*-acting ribozyme to create *trans*-acting ribozymes which are capable of cleaving external strands of RNA. This principle of converting a small ribozyme into two parts, an enzyme portion that cleave the RNA substrate portion at specific sites, opened up the possibility of using ribozymes to cleave an RNA sequence corresponding to defective

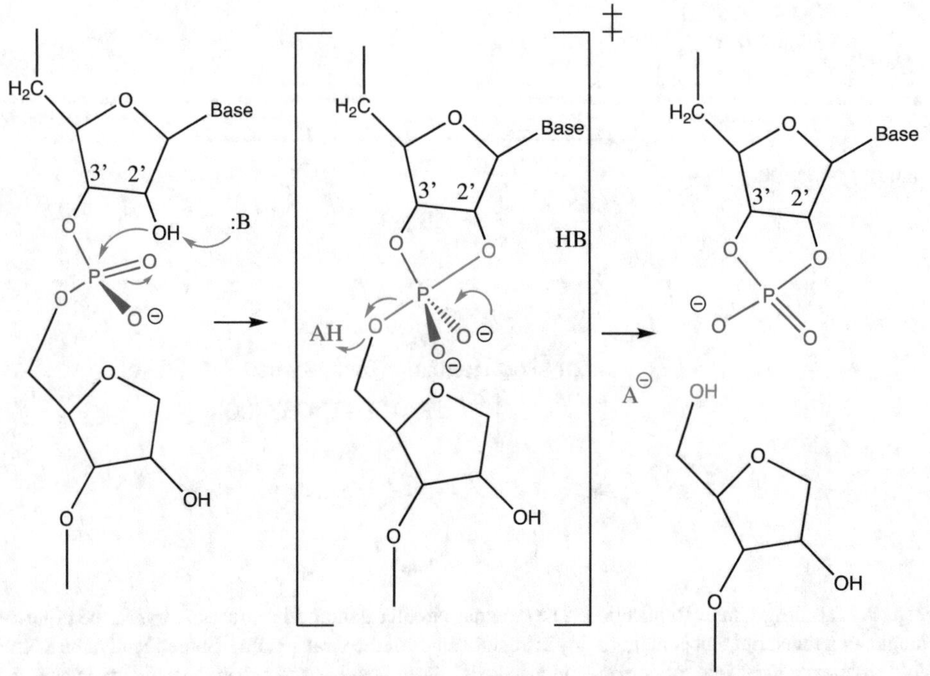

Fig. 3.2 A Mechanism of RNA Backbone Cleavage. The nucleophilic 2'-OH is activated by a general base and to act on the 3' phosphate. In the transition state, the 2'-OH and the leaving phosphate have to be in-line and the phosphate has to adapt a trigonal bipyramidal configuration. The leaving group picks up a proton resulting in a 5'-OH and a cyclic 2',3'-cyclic phosphate. Often magnesium hydroxide $[L_5Mg(OH)]^+$ act as the base (where, L is water or RNA ligands). The ligands bound to Mg^{2+} and the local RNA environment may alter the pKa of the magnesium hydroxide

genes or those of an infectious virus, like the coronavirus. Much research in small biotechnology companies is directed toward use of ribozymes for therapy. This idea of RNA as molecular scissors is useful in the laboratory too, for example, to generate a specific RNA transcript.

Backbone cleavage of RNA is carried out using a S_N2 mechanism. The 2'-OH of a ribose acts upon the 3' phosphate, and the neighboring nucleotide acts as a leaving group (Fig. 3.2). The 2'-OH is not a good nucleophile and an alkoxide is not a good leaving group. For a nucleophile to form, the 2'-OH needs to be converted into a 2'-O$^-$ by deprotonation. The base that deprotonates the 2'-OH varies between different ribozymes. Often it is a hydroxide ion coordinated to a magnesium ion that acts as the base. A nucleotide base in the active site can also perform the role of a general base. The leaving group is stabilized by a general acid, which can be a neighboring nucleotide or a water molecule coordinated to a magnesium ion. In order to further lower the energy of the transition state, and by extension the activation energy, the doubly charged phosphate is stabilized by positive charges, often carried by a divalent cation like magnesium ions.

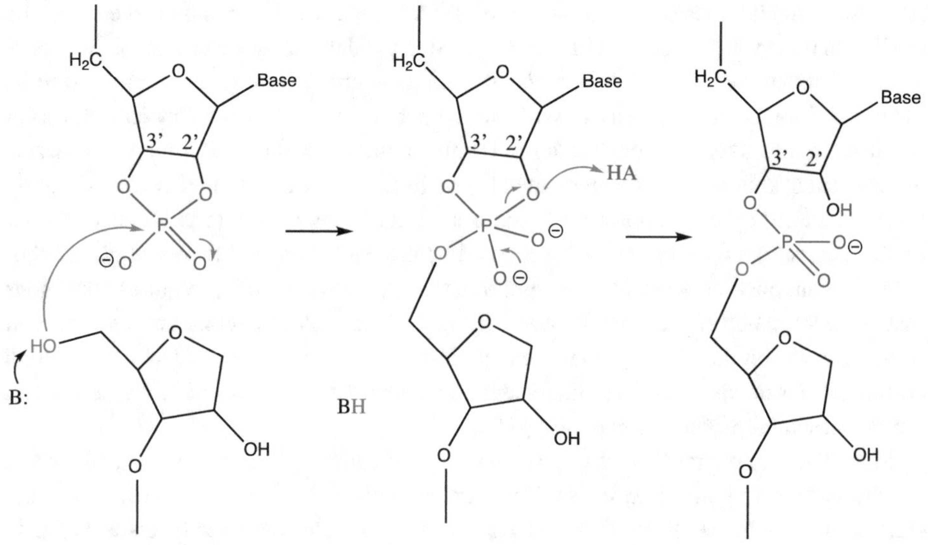

Fig. 3.3 The Backbone Ligation Reaction Mechanism. The ligation reaction catalyzed by small ribozymes shows 5'-OH is deprotonated to act on the cyclic phosphate. This results in the cyclic 2',3'-phosphate to open and restore the 2'-OH on the ribose. Here, B represents any general base and HA represents any general acid; magnesium bound ligands (often water molecules) are often seen to perform this role

The products of the cleavage reaction are one strand ending in a 2',3'-cyclic phosphate and the second strand ending in a 5' hydroxyl terminus.

The ligation reaction is the reverse of cleavage reaction (Fig. 3.3).

3.3 The Role of Metal Ions

Metal ions are crucial to the functioning of all small catalytic RNA directly or indirectly [3–9]. RNA molecules require cations to neutralize the negative charges on the backbone phosphates to fold into secondary and tertiary structures. Small catalytic RNA are no exception. Under physiological conditions, the backbone charge neutralization is likely accomplished by K^+ and Mg^{2+}; in vitro molar concentrations of monovalent ions, such as Na^+ or Li^+, can accomplish charge neutralization and enable RNA folding and catalysis for certain ribozymes. Although these serve as excellent proof-of-principle experiments, RNA's ability to form alternate structures under these very different ionic conditions, and the associated alteration of base pKas, leave physiological relevance of these experiments questionable while providing intriguing insights into RNA dynamics.

The role of metal ions in catalysis and structure formation is difficult to study as metal ions serve to both neutralize negative charges (diffuse binding) and interact in a site specific manner [12–16]. The magnesium ions are also spectroscopically silent therefore high resolution structural data (resolution of ~2 Å) are necessary to ascertain their specific binding in the crystal structures [12]. To conclusively identify magnesium ion binding sites in crystal structures, six ligands in an octahedral geometry around magnesium need to be identified. Most structural data do not have the necessary resolution to conclusive assign magnesium ions over sodium or water (all with similar numbers of electrons); and hence, the particular assignment of lighter metal ions in the structures should be taken lightly unless validated by other approaches. The heavier metal ion are easier to identify (greater electron density), however, the sites occupied by these ions may not represent physiologically relevant positioning of ions. In addition, the crystallization often requires conditions that are often much higher in salts and may favor local conformations that are different from those in solution. Indeed, experimental in vitro conditions (use of Na^+ ions or high concentrations of Mg^{2+} ions) are likely stabilizing alternate structures of RNA, as seen by in vivo structural probing of genomic RNA.

Many small ribozymes directly use cations, specifically Mg^{2+}, in catalysis, while others use nucleotides as general acid/base. There are a number of different roles which magnesium ions play in catalysis. A water ligand or hydroxide ion directly coordinated to magnesium ion could act as a general acid or base. A metal ion could also act as a Lewis acid itself by forming an inner-sphere complex with the RNA. A positively charged ion often stabilizes the negative charge developing in the transition state and it stabilizes the leaving group. Metal ions also help in the formation of structures necessary for catalysis and are likely to play a role in altering the pKa of the neighboring nucleotides and sugars.

Metal Ions in Structures A number of different techniques are utilized to examine the role of metal ions in catalysis. One way to gain insight into the function of metal ions is by determining where metal ions are bound to RNA molecules. The idea here is that the location of metal ions indicates higher occupancy, and lower mobility, of ions in these sites, thus providing clues to their role in catalysis. As high resolution crystal structures of a number of small catalytic RNA molecules have been solved, many coordinated metal ions, often manganese-bound sites, have been identified. NMR experiments have also been able to identify divalent cations binding sites (Fig. 3.4). Each method has its limitations. There are additional issues that are important to consider including, metal ions may localize differently in vivo, and their location may change throughout the reaction (dynamic rearrangements with shifting electrical fields) or may involve transient interactions within a network of ions that are difficult to isolate or study.

Thiophosphate "Rescue" Experiment A common technique for studying the role of Mg^{2+} at a particular site is to substitute a specific non-bridging oxygen atom (pro-R or pro-S) on the phosphate with a sulfur atom. If the sulfur atom replaces the oxygen that binds the catalytic magnesium ion, then catalytic reaction will stop due to the different ion binding properties of oxygen versus sulfur.

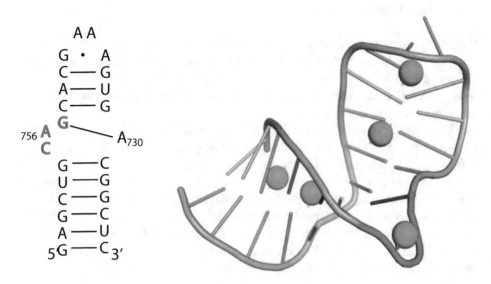

Fig. 3.4 Magnesium Binding in Domain VI of VS Ribozyme. (**a**) Secondary structure of the stem-loop derived from domain VI. (**b**) Five magnesium ions (green balls) are seen in the NMR structure of the small RNA derived from Domain VI; A756, which acts as a general acid in catalysis is extruded into the minor groove by the formation of the S-turn. The negative charges in the S-turn are stabilized by magnesium ions

The magnesium ions (smaller, harder) are expected to only bind hard ligands (oxygens and nitrogens), and hence, are not expected to bind to the sulfur atom. When larger metal ions (larger, softer), like Cd^{2+}, are added to the solution, these will preferentially bind softer ligands like sulfur, and are expected to contribute to the reaction in a manner similar to magnesium ions. Therefore, addition of softer metal ions will "rescue" catalysis. The thiophosphate experiments (in theory) will only show cleavage at the expected sites and only in the presence of thiophilic metal ions. The reality of sulfur substitutions is a bit more complicated but the principles on which thiophosphate rescue are based are generally applicable and have been used to study the catalytic role of magnesium ions in large and small catalytic RNA. We will encounter a few of these experiments in our discussions of RNA structures and functions.

The rescue technique is expected to differentiate between the role divalent ions have in RNA folding from that in catalysis. In a rescue experiment, a non-bridging oxygen on a specific phosphate group is replaced with a sulfur atom. Magnesium ions are present in solution, allowing RNA to fold but are not expected to coordinate with the sulfur atom in order to participate in the catalysis reaction. Catalytic rate is then measured in the presence of Mg^{2+} and with and without added Mn^{2+} or Cd^{2+}. If the phosphate of interest requires a coordinated divalent ion for catalysis, and not just for folding, the catalytic rate should increase in the presence of thiophilic metal ion in a pH-dependent manner (corresponding to the pKa of the metal-bound hydroxide participating in the reaction.)

3.4   Regulation

Most small ribozymes in nature are active as soon as they are transcribed. Some like GlmS ribozyme will require cofactor binding in order to become a ribozyme, thus regulating the activity of the ribozyme. The principles of regulation of activity of the ribozyme will become clearer as we learn more about different catalytic RNA and their functions via comparative genomic analyses.

The principles that we have learned from studying small catalytic RNA are being applied to create artificial RNA enzymes. Often artificial enzymes are designed by modifying a known ribozyme in vitro. For example, a modified hairpin ribozyme (discussed below) has been constructed to fold into a catalytically competent ribozyme only upon binding to a specific oligonucleotide [17]. Modified hairpin ribozymes have also been created that are active when bound to an oxidized FMN but inactive when bound to a reduced FMN, leading to a ribozyme whose activity could be turned up or down with electrochemical redox [17]. Thus, application of ribozyme chemistry is a vibrant area of research and many interesting applications are being explored.

3.5   Examples of Four Small Catalytic RNA

Small catalytic RNA share many common features. They all catalyze a similar backbone cleavage reaction, they all require metal ions in order to fold and function, and they all accomplish catalysis without any protein involvement. Many have evolved independently. They do, however, vary in significant ways. A few examples of small ribozymes are discussed below to illustrate the diversity of RNA structures involved in catalysis.

Hepatitis Delta Virus (HDV) Ribozyme Hepatitis is a human disease that leads to increased chances of liver failure and liver cancer. It is caused by a small virus-like particle called hepatitis delta virus (HDV). Hepatitis D virus is a satellite virus of the Hepatitis B virus.

The HDV has a 1.7 kb genome and replicates by a rolling circle mechanism. The HDV ribozyme is a small catalytic RNA isolated from the Hepatitis D Virus [18–22]. The viral genome is transcribed multiple times into a single large multimeric RNA strand which is subsequently cleaved into individual monomers by the HDV ribozyme. HDV ribozyme is found in both genomic and antigenomic strands of the virus. HDV-like ribozymes have been found in all kingdoms of life, including the human genome. The CPEB3 gene in humans contains a CPEB3 ribozyme which is HDV-like; this gene is known to play a role in memory. Understanding the structure and function of the small ribozymes has direct implications for human health.

The HDV catalytic motif is ~85 nucleotides long [18]. This relatively short sequence folds into an incredibly stable structure which can remain intact in relatively extreme conditions, such as 80 °C heat or solvent containing 5 M urea or formamide. The HDV ribozyme structure is a compact structure containing five base-paired regions (paired regions are marked as P1, P2, P3, P4, and P1.1) that form a nested double, pseudoknot structure. The five helical segments form two stacks: P1, P1.1, P4 stack nearly coaxially; P2 and P3 form a second coaxial stack. The cleavage site is on the 5' end; there are no specific sequence requirements on the 5' of the cleavage site—it just requires the presence of a nucleotide; this is an unusual feature of this ribozyme. A key requirement for cleavage reaction is a G•U base pair in position +1 (the first base pair of P1 helix). It also requires a specific cytosine, C75, located in junction between helix P4 and P2 (J4/2). C75 is located in the active site near the 5' hydroxyl and participates in general acid–base reaction.

The active site pocket is in the middle of the ribozyme and is very electronegative. A twist in the backbone of J4/2 positions (purple in Fig. 3.5) a cytosine, C75, at the bottom of the active site pocket [18]. The amino group of this cytosine forms a hydrogen bond with the non-bridging oxygen of another cytosine nucleotide, C22.

The HDV ribozyme accomplishes catalysis by using the C75 nucleobase as a general acid. While free RNA nucleobases do not have functional groups with pKas near the physiological pH, local electronic environments within RNA likely induce shifts in pKas [22]. In the case of HDV, the electronegative environment of the binding pocket, along with the hydrogen bond between cytosine nucleotides, raises the pKa of the C75 N3 to be closer to the physiological pH. The amine is then able to act as an acid and protonate the leaving group. A metal ion, which is often magnesium ion, deprotonates the 2'-OH to make it into a nucleophile. Together, C75 and the metal ion enable the ribozyme to catalyze a cleavage reaction (Fig. 3.6).

The HDV ribozyme has been found to only catalyze the cleavage reaction, and does not perform the reverse ligation reaction. While ligation is a critical step of rolling circle replication, a different cellular enzyme may perform the ligation of the monomers.

Hairpin Ribozyme The hairpin ribozyme was discovered in the tobacco ringspot virus [22, 23]. Similar to the HDV ribozyme, the hairpin ribozyme cleaves a long strand of RNA containing multiple copies of a viral genome into the individual monomers. The hairpin ribozyme is a well-studied small catalytic RNA molecules, with an extensive structural and catalytic information available.

The hairpin ribozyme is one of the smallest catalytic RNA sequences. It is composed of ~50 nucleotides [24]. The hairpin ribozyme consists of two independently folding domains (A and B); each domain is composed of two helices that are separated by an internal loop [11].

A fully folded hairpin ribozyme bends in such a way that the two internal loops (loop A and B) dock. These docked internal loops form the catalytic core of the ribozyme (Figs. 3.7 and 3.8) [4–8, 11]. The adenine preceding the cleavage site is stacked in the stem and the guanine +1 is extruded from the helix to be accommodated in a complementary pocket. The docked conformation is stabilized by two major interactions: a Watson–Crick base pair

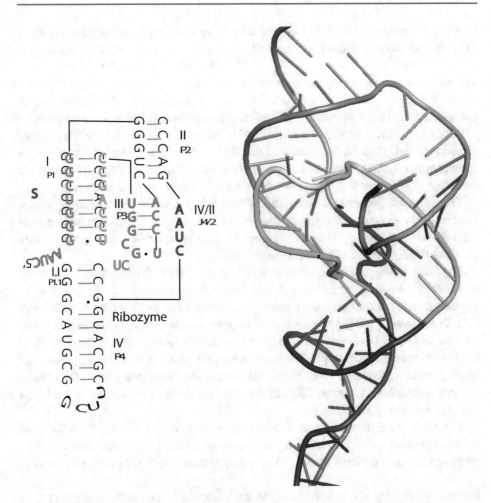

Fig. 3.5 Hepatitis D Virus Ribozyme. The secondary and tertiary structure of HDV ribozyme are shown. The crystal structure of the HDV ribozyme (2.4 Å resolution) shows two pseudoknotted regions (colored) in the middle of the structure. The active site of HDV ribozyme is buried in the middle of the structure; the 5′ end of the substrate strand is colored in red. The figure made with PDB file 4PRF using PyMol

between a loop A guanine and a loop B cytosine, and a ribose zipper between the two loops. The docking step is necessary for the cleavage reaction to occur; it requires the rearrangement of loops A and B as they come into close proximity. Residues in the loops (particularly, A38 in loop B and G8 in loop A) play an important role in stabilizing the transition state. The overall reaction process consists of at least few distinct steps: docking of the loops, cleavage, undocking and dissociation. HDV forms a four-way junction structure (Figure 3.7b). These steps have been studied using fluorescence resonance electron transfer (FRET) analysis where the distance between two dyes can be

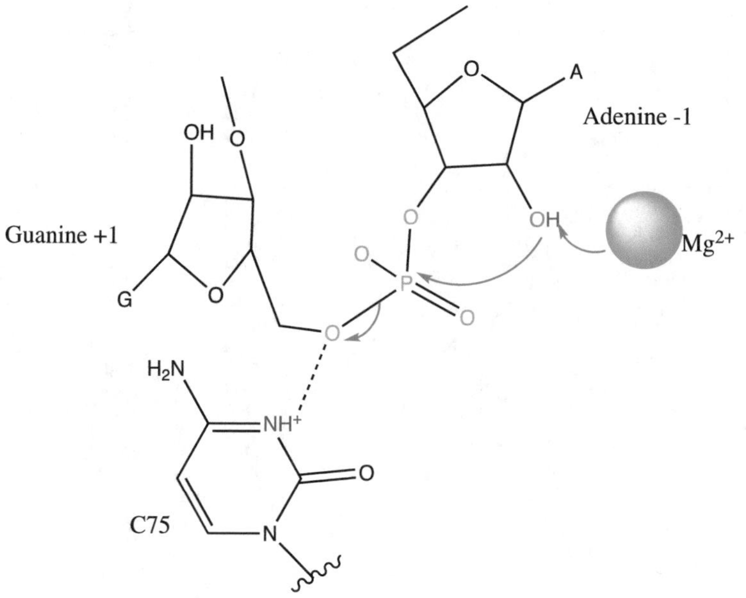

Fig. 3.6 Catalytic Mechanism of HDV Ribozyme. Magnesium ions are bound to six ligands. [L$_5$Mg (OH)]$^{2+}$, where L can be water or RNA ligands. The magnesium ion bound hydroxide deprotonates the adenine 2'-OH; C75 nucleobase acts as a general acid to facilitate the reaction

measured to determine the bend angles. The steps proposed for the hairpin are shown schematically in Fig. 3.8.

The crystal structure of the four-helix bundle was solved using a 2'-methoxy substituent (in the place of 2' hydroxyl) in order to prevent the cleavage reaction. The four helices radiate from a perfectly base-paired four-way junction (Figure 3.7b). Helices form two coaxial stacks: I with II, and III with IV. The two stacks cross at ~60° angle. The junction allows the central portions of helices I and IV to dock through their minor grooves in order to form the active site. Interestingly, no bound metal ions are seen in the junction. The hydroxyl radical footprinting experiments found the core to be inaccessible by solvents. Well-ordered metal ions are seen in loop A and are further supported by FRET experiments. A S-turn, ribose zipper, a cross-strand puring stack are seen in the structure (highlighted in Figure 3.7d). Two key points to note here are: first, large rearrangements occur in the RNA to form an active site. Second, the loop A and B regions are largely base paired; the interactions between the two loops are important for the catalytic activity.

The exact sequences within the paired regions have not been found to significantly alter the catalytic activity so long as they support the formation of the helices. The conserved nucleotides are located within the internal loops; a guanine 3' of the cleavage site (+1 site) has been found to be required for catalytic activity. Any nucleotide can be found at the −1 position (5' position of the cleavage site in loop A). The N3 of adenine at this position shows a single hydrogen bond with A9. This base pair geometry is compatible with all four

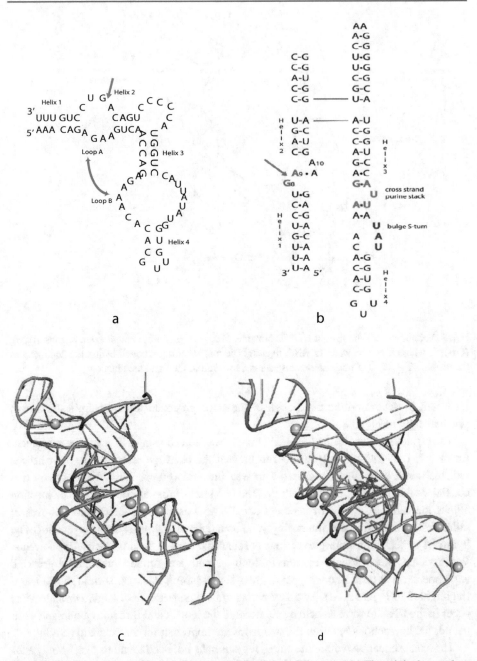

Fig. 3.7 The Secondary and Tertiary Structures of the Hairpin Ribozyme. (**a**) The original secondary structure of the RNA was first determined using the thermodynamic structure prediction programs. (**b**) A corrected secondary structure that corresponds to the tertiary structure. (**c**) Tertiary structures for hairpin ribozyme are shown; pink spheres represent bound calcium ions. (**d**) The active site is in the center of the molecule. The residues in loop A, A8 and G9 (red), the S-turn containing A38 in loop B (purple), and the cross purine stack region (green) are highlighted. Figure (d) is a slightly rotated version of figure (c). Figure made using PDB file 1M5K in PyMol

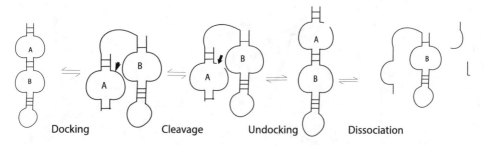

Fig. 3.8 Various Conformations of the Hairpin Ribozyme. A schematic of different conformations that are proposed to occur between loop A and B in the hairpin ribozyme [24]

bases as purines have N3 and pyrimidines O2 atom in the equivalent positions. The geometry of this base pair causes it to cross strand stack with G8. This, along with conformation of G +1 causes broadening of the minor groove.

The mechanism by which the hairpin ribozyme accomplishes catalysis is thought to be similar to the HDV ribozyme. Metal ions and other catalytic cofactors are not thought to be directly involved in catalysis, though metal ions are necessary to enable proper folding [4–8]. An adenine and guanine located in the active site function as general acids and bases to activate the leaving group and nucleophile (Fig. 3.9).

Whereas the HDV ribozyme has only been found to catalyze the cleavage reaction, the hairpin ribozyme has been found to favor the ligation reaction over the cleavage reaction. In rolling circle replication, after monomers are cleaved, each monomer must cyclize back into a circle so it can be transcribed and multiply further. The hairpin ribozyme likely has evolved to efficiently catalyze this ligation step.

Hammerhead Ribozyme   The hammerhead ribozyme is similar to the hairpin ribozyme in a number of ways. It was discovered in 1987 within RNA satellite viruses which infects plants, and it utilizes the rolling circle replication mechanism [25]. It is one of the most well-studied small catalytic RNA. The hammerhead ribozyme crystal structure was the first catalytic RNA structure to be solved in 1995 [26, 27]. Despite this, the structure-function relationship of the hammerhead ribozyme and its catalytic mechanism are still debated [4–8].

The hammerhead ribozyme sequence is relatively short and forms a simple structure (Fig. 3.10). Three helices form a Y-shape, with stems II and III coaxial and stem I at an angle from this axis. The active site pocket is at the junction of these three helices. While the exact lengths and content of the three stems can vary, there are 11 highly conserved nucleotides in the active site pocket.

Several secondary structures feature outside of the catalytic core are crucial for efficient cleavage. The hairpin loop at the end of helix II and a bulge within helix I interact to enable the ribozyme to fold for an efficient reaction.

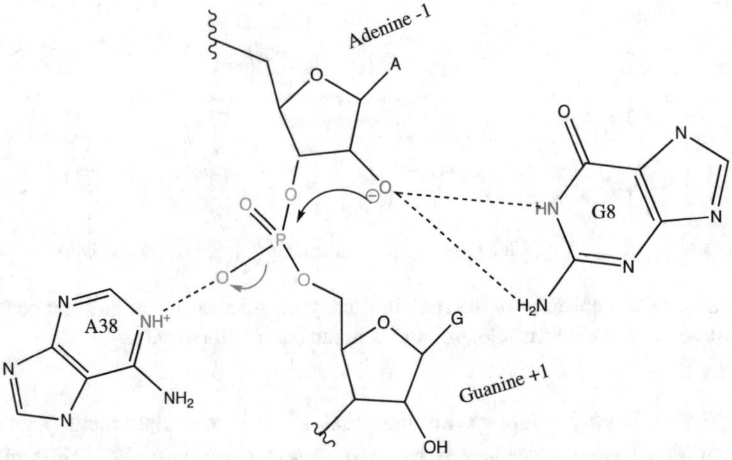

Fig. 3.9 Cleavage Mechanism for the Hairpin Ribozyme. This mechanism shows the role of the nucleotides G8 and A38 as general base and acid in the reaction mechanism for cleavage

Like the hairpin ribozyme, the hammerhead ribozyme is also thought to undergo important conformational changes in the process of cleavage. The crystalized structure of the ground state hammerhead ribozyme does not have the 2' OH nucleophile positioned as it needs to be in order to attack the neighboring phosphate. This suggests that the ribozyme changes conformation briefly in order to position these groups so that the reaction can proceed.

A magnesium ion bound to A9 is likely involved in catalysis. It is proposed to bridge the negative charges of the approaching phosphates in the transition state. In a Mn^{2+} bound crystal structure of the hammerhead ribozyme at 2 Å resolution, five Mn^{2+} binding sites were identified, including one in the active site. Despite years of research into the role of metal ion, G8, and G12 in the cleavage mechanism, the role of each in the transition state remain unclear. To propose an accurate mechanism, the structural data has to match the biochemical data. In the crystal structure of the hammerhead ribozyme, positions of the key residues and their distance from each other is such that multiple different mechanisms can be proposed (Fig. 3.11). The biochemical data matches the one-metal experiments best but the binding site of the metal ion expected to assist in catalysis is further from the cleavage site, at A9. It is possible that under certain conditions, hammerhead could function in a no-metal ion environment as catalysis is possible in very high monovalent salt conditions. The thio-rescue experiments support the involvement of metal ions in the mechanism.

***GlmS* Ribozyme** The *glmS* motif was discovered in the 5' UTR of a gene encoding for glucosamine-6-phosphate (GlcN6P) synthetase. The motif binds GlcN6P, which catalyzes a *cis*-cleavage reaction [28–31]. The product of this cleavage, the mRNA, is targeted by cellular RNases for rapid degradation, effectively downregulating translation of the

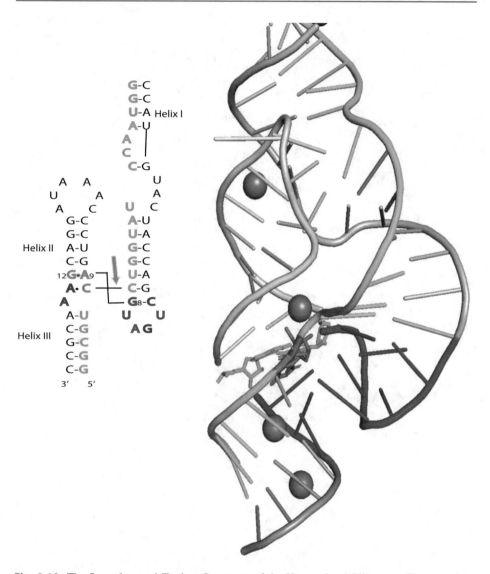

Fig. 3.10 The Secondary and Tertiary Structures of the Hammerhead Ribozyme. The secondary structure and tertiary structure of hammerhead ribozyme with substrate strand (cyan) and the cleavage site (red) color coded. The crystal structure (2.0 Å resolution) is shown with conserved residues highlighted (dark blue); The active site A9•G12 base pair that binds to an essential magnesium ion is shown as sticks (green). The sphere represent the bound manganese ions. Figure drawn using PDB file 2OEU in PyMol

GlcN6P synthetase transcript. As such, the *glmS* motif can be considered both a small catalytic RNA and a riboswitch (Chap. 7, Riboswitch). Although, as a riboswitch, it doesn't

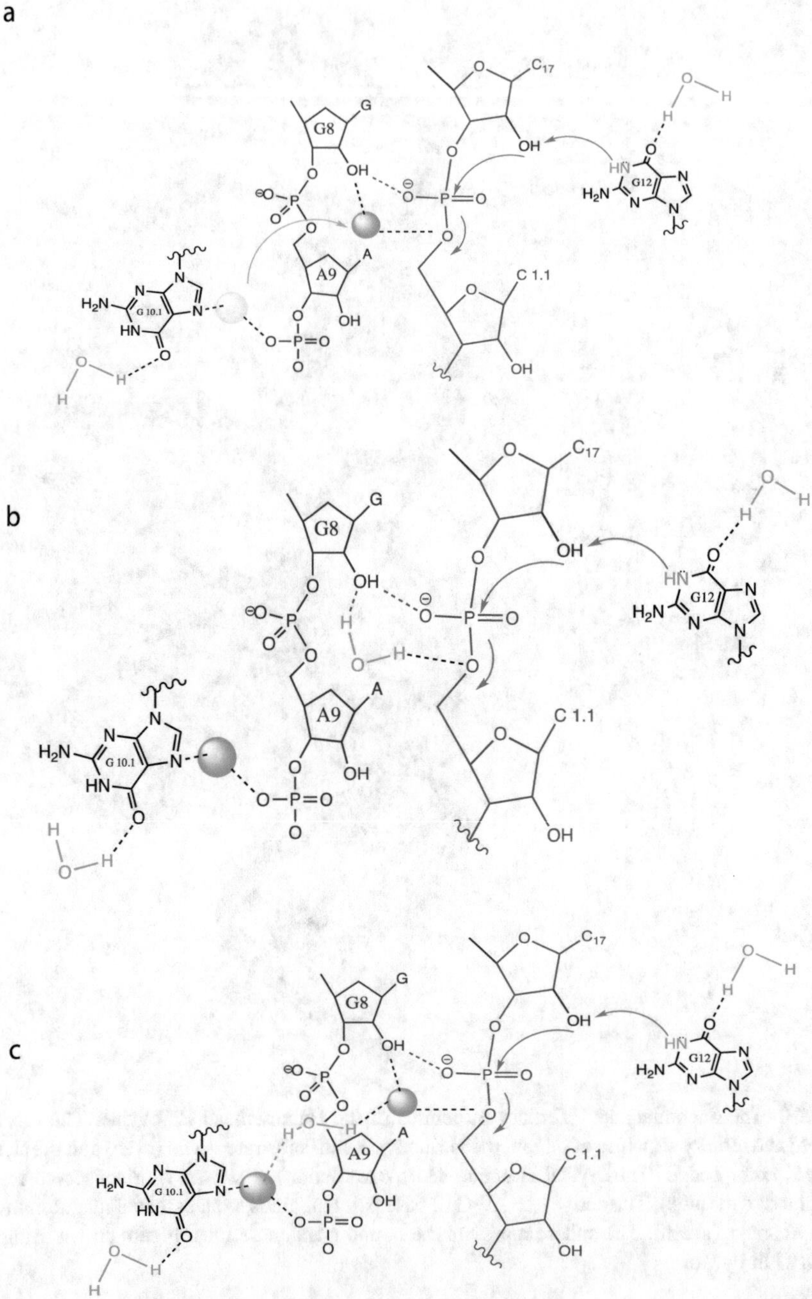

Fig. 3.11 Three Potential Mechanisms for Hammerhead Catalysis. The RNA chain with the cleavage site is shown in blue; the site of metal ion binding and nucleotides implicated in catalysis are shown in purple and black. (**a**) The metal ion moves from the original location at A9 to be part of the reaction mechanism during the transition state in this one-metal mechanism; (**b**) Alternatively, a second metal ion is recruited into the active site in the two-metal mechanism; the two metal ions

undergo conformational changes upon effector binding. Since the discovery of glmS ribozyme, other metabolite-binding ribozymes have been discovered [29].

The motif is roughly 120 nucleotides long and folds into a structure containing 8 small stems which stack coaxially (Fig. 3.12). In the crystal structure, stems P1, P2, P3, P3.1 form one helical stack. A second helical stack formed by P4 and P4.1 [31]. One pseudoknot, P2.1, is formed between P1, P2, and 5′ end of P2 loop. The second pseudoknot, P2.2, is formed by base pairing interactions of the 5′ end of the ribozyme and the 3′ side of P2 loop. The double pseudoknots, P2.1 and P2.2, form the catalytic core that includes the GlcN6P binding pocket [31]. The P2.1 and P2.2 pseudoknots form the sides of an active side that is capped by a single base pair A46 and U51 (closing base pair of P2 loop). A28 stacks directly on G1 (3′ of the scissile bond), which in turn is stacked on top of GlcN6P. For catalysis, the P3 and P4 helices are dispensable.

Binding of GlcN6P is achieved through recognition of its phosphate and sugar moieties. Two magnesium ions bind the phosphate group. The glucosamine ring makes direct interactions with the RNA. Both magnesium ions are positioned in the major groove of the P2.2 helix in the effector binding pocket. One magnesium makes additional contacts with A28, G53, and G54 of the RNA. An deazaadenosine substitution in the ribozyme interferes with the activity of the ribozyme, supporting a key role for A28. The active site itself is devoid of metal ions and hence, GlmS ribozyme is expected to use nucleotides for general acid/base reactions. In vitro, a high concentration of monovalent ions support catalytic activity in the absence of magnesium ions. The scissile phosphate and flanking nucleotides are properly oriented in the active site. The nucleotide at 5′ position of the scissile bond, A-1, is oriented by interactions with G57. Mutation of G57, G57C, or G57A, abolish nearly all activity.

The catalytic core contains a rigid binding pocket which can recognize and bind GlcN6P through the sugar moiety and the phosphate (Fig. 3.13). Notably, the conformation of the *glmS* ribozyme has been found to not change upon binding of GlcN6P. This is unusual, as traditional riboswitches change conformation significantly upon binding of their cofactor (also see the Chap. 7, riboswtiches).

The mechanistic pathway through which *glmS* catalyzes reactions is rare amongst the ribozymes [28–31]. Like the HDV and hairpin ribozymes, a nucleobase is thought to play a significant catalytic role. The N1 of the nucleotide G33 is positioned optimally to activate the nucleophilic 2′ OH (Cochrane). The defining feature of *glmS* catalyzed cleavage, however, is the participation of GlcN6P in the reaction mechanism. The amine group on

Fig. 3.11 (continued) interact via a water or a hydroxide to participate in the cleavage and stabilization of the leaving group. (**c**) In a no-metal mechanism, the metal ions are not involved in the catalysis reaction itself but play a role in forming the correct structure of the RNA to position the correct nucleotides in the cleavage site. The G12 nucleotide acts as a general base and G8 acts as a general acid—either directly or through water mediated interactions with the cleavage site

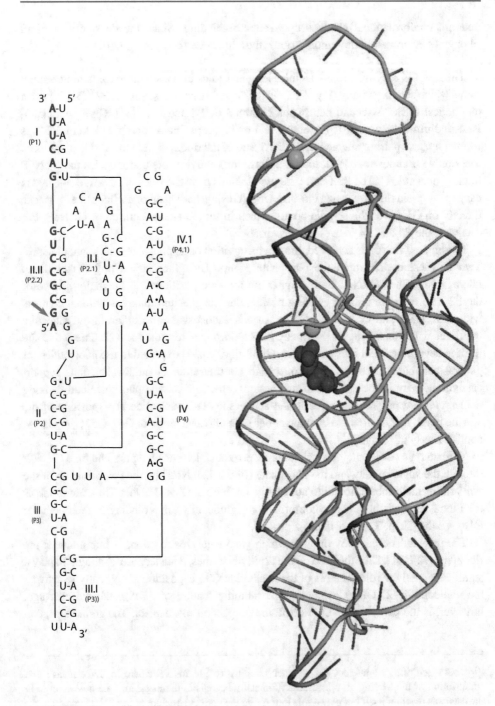

Fig. 3.12 The Secondary and Tertiary Structures of the *glmS* Ribozyme. The glmS ribozyme binds to GlcN6P phosphate (blue) to perform the catalysis reaction. The substrate strand is shown in magenta. The 5′-end of the substrate strand is buried inside the ribozyme and is near the GlcN6P binding site. The figure was made using PDB File 3G8T (3 Å resolution) using PyMol

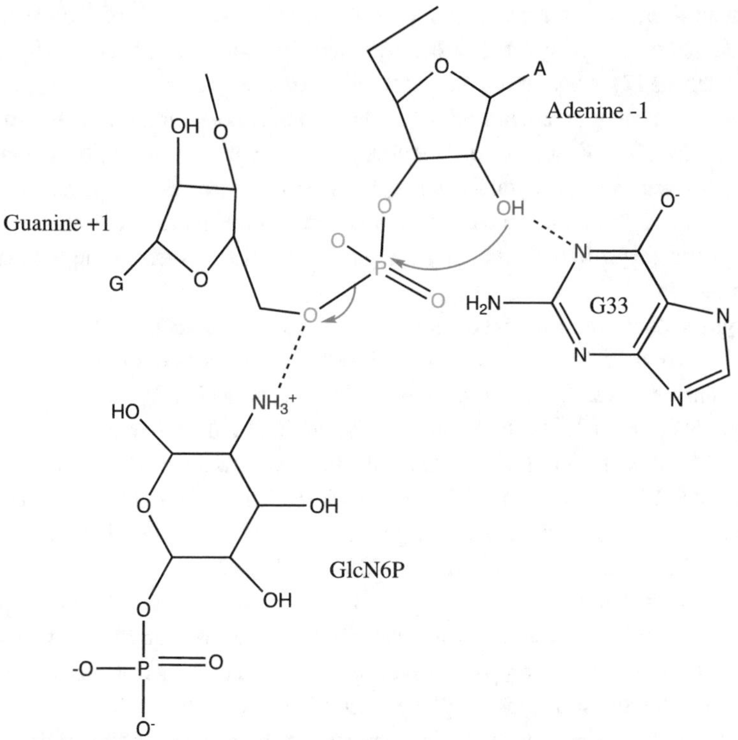

Fig. 3.13 A Proposed Mechanism for *glmS* Catalyzed Backbone Cleavage. The *glmS* motif uses both a nucleobase (G33) and a cofactor (GlcN6P) for catalysis. Under certain conditions, the reaction could be supported by magnesium ions in the absence of the cofactor

GlcN6P has a pKa of around 8.2, while the G33 N1 has pKa between 9.3 and 10.4. making it well suited to act as a general acid/base around physiological pH. Ideally, the pKas of general acid/base groups should be close to neutral and separated by less than 2 pH units to exist in both protonated and deprotonated states at physiological pH. The amine of GlcN6P is expected to donate a proton to the leaving group, thus helping stabilize the leaving group and provide substantially more catalytic power than G33, at physiological pH.

The GlcN6P cofactor, upon binding, is used directly in catalysis. This use of organic molecules by ribozymes further supports the hypothesis that RNA was the original genetic material (the RNA World hypothesis) and crucially it provides a potential path to a larger repertoire of RNA catalyzed reactions that are yet to be discovered.

3.6   Implication for Health and Medicine

Many of the small RNA-based ribozymes are derived from pathogenic viruses. Understanding the structures and cleavage mechanisms of these ribozymes will help in efforts to develop new antiviral medicines [32–36]. The HDV ribozyme, hairpin ribozyme, hammerhead ribozyme, and *glmS* ribozyme are just four examples of the small catalytic RNA discussed here—many others are known and even more have been identified through genomic analysis [7]. Many more are yet to be discovered.

All ribozymes catalyze reactions of a nucleic acid backbone through a $2'$ hydroxyl that is activated to act on a $3'$ phosphodiester linkage, or favor the reverse backbone ligation reaction. Different small catalytic RNA may use metal ions in different ways, but all require metal ions in order to fold into a catalytically competent molecule. Some can even perform these reactions in profoundly nonphysiological conditions, such as high salts or high temperatures.

Ribozymes that were self-cleaving (i.e., perform cis-cleavage) were artificially converted into two strands—one strand that is cut is the substrate strand and the remainder of the RNA acts as an enzyme (ribozyme)—it allowed *trans-cleav*age of RNA. This opened up possibilities to cut RNA with precision, whether it be a defective gene or a pathogenic nucleic acid from bacteria or a virus. Due to the transient nature of RNA in the cell, the side effects of the RNA therapy are expected to be minimal. The ideas developed in examining small catalytic RNA were also directly applicable in developing CRISPR-Cas systems that use guide RNA-based targeted cleavage (CRISPR, Chap. 9).

Understanding the biochemistry of small ribozymes has led to developing new RNA-based technologies. For example, hammerhead ribozyme are routinely incorporated into transcription reactions to produce precise ends of the transcript. As we learn about diversity of catalytic strategies employed by RNA, we will develop new antibiotics precisely tailored to viral and bacterial genomes. We are also finding RNA sequences that form ribozyme structures in the human genome. Synthetic ribozymes are being created using the principles of RNA recognition and cleavage to treat defective or pathogenic RNA and DNA [29].

Take Home Message
- Short sequences of RNA have evolved to catalyze reactions of RNA cleavage and ligation using complex structural features.
- Exact role of metal ions varies between ribozymes, but all require metal ions for catalysis, either directly or indirectly.
- The HDV ribozyme, hairpin ribozyme, hammerhead ribozyme, and *glmS* ribozyme are examples of small catalytic RNA. The principles of RNA structures, conformations, kinetics, and thermodynamics can be learned by studying small catalytic RNA.

References

1. Cech TR, Zaug AJ, Grabowski PJ. In vitro splicing of the ribosomal RNA precursor of Tetrahymena: involvement of a guanosine nucleotide in the excision of the intervening sequence. Cell. 1981;27:487–96.
2. Guerrier-Takada C, Gardiner K, Marsh T, Pace N, Altman S. The RNA moiety of ribonuclease P is the catalytic subunit of the enzyme. Cell. 1983;35:849–57.
3. Diegelman AM, Kool ET. Generation of Circular RNAs and Trans-Cleaving Catalytic RNAs by Rolling Transcription of Circular DNA Oligonucleotides Encoding Hairpin Ribozymes. Nucleic Acids Res. 1998;26:3235–41.
4. Doherty EA, Doudna JA. Ribozyme structures and mechanisms. Ann Rev Biochem. 2000;69:597–615.
5. Fedor MJ, Williamson JR. The catalytic diversity of RNAs. Nat Rev Mol Cell Biol. 2005;6:399–412.
6. Carola C, Eckstein F. Nucleic Acid Enzymes. Curr Opin Chem Biol. 1999;3:274–83.
7. Weinberg CE, Weinberg Z, Hammann C. Novel ribozymes: discovery, catalytic mechanisms, and the quest to understand biological function. Nucleic Acids Res. 2019;47:9480–94.
8. Lilley DM. The origins of RNA catalysis in ribozymes. Trends Biochem Sci. 2003;28:495–501.
9. Barrick JE, Corbino KA, Winkler WC, et al. New RNA motifs suggest an expanded scope for riboswitches in bacterial genetic control. PNAS. 2004;101:6421–6.
10. Cochrane JC, Lipchock SV, Strobel SA. Structural investigation of the GlmS ribozyme bound to its catalytic cofactor. Chem Biol. 2007;14:97–105.
11. Reymond C, Beaudoin JD, Perreault JP. Modulating RNA Structure and Catalysis: Lessons from Small Cleaving Ribozymes. Cell Mol Life Sci CMLS. 2009;66:3937–50.
12. Auffinger P, Grover N, Westhof E. Metal Ion Binding to RNA 9:1–35. In: Siegel, Siegel, Siegel, editors. Metal Ions in Life Science Biology Series. Oxford: Oxford University Press; 2011.
13. Bonneau E, Legault P. NMR localization of divalent cations at the active site of the neurospora vs ribozyme provides insights into RNA-metal-ion interactions. Biochemistry. 2014;53:579–90.
14. Auffinger P, Ennifar E, D'Ascenzo L. Deflating the RNA Mg^{2+} bubble: stereochemistry to the rescue! RNA. 2021;27:243–52.
15. Zheng H, Shabalin IG, Handing KB, et al. Magnesium-binding architectures in RNA crystal structures: validation, binding preferences, classification and motif detection. Nucleic Acids Res. 2015;43:3789–801.
16. Fischer NM, Polêto MD, Steuer J, van der Spoel D. Influence of Na + and Mg2+ ions on RNA structures studied with molecular dynamics simulations. Nucleic Acids Res. 2018;46:4872–82.
17. Müller S, Appel B, Krellenberg T, et al. The many faces of the hairpin ribozyme: structural and functional variants of a small catalytic RNA. IUBMB Life. 2012;64:36–47.
18. Shih IH, Been MB. Catalytic Strategies of the Hepatitis Delta Virus Ribozymes. Annu Rev Biochem. 2002;71:887–917.
19. Chen JH, Yajima R, Chadalavada DM, Chase E, Bevilacqua PC, Golden BL. A 1.9 A crystal structure of the HDV ribozyme precleavage suggests both Lewis acid and general acid mechanisms contribute to phosphodiester cleavage. Biochemistry. 2010;49:6508–18.
20. Kapral G, Jain S, Neoske J, Doudna JA, et al. New tools provide a second look at HDV ribozyme structure, dynamics and cleavage. Nucl Acid Res. 2014;42:12833–46.
21. Ferré-d'Amaré AR, Zhou K, Doudna JA. Crystal structure of a hepatitis delta virus ribozyme. Nature. 1998;395:567–74.
22. Strobel SA, Cochrane JC. RNA Catalysis: Ribozymes, Ribosomes, and Riboswitches. Curr Opin Chem Biol. 2007;11:636–43.

23. Rupert PB, FerreÂ-D'Amare AR. Crystal structure of a hairpin ribozyme±inhibitor complex with implications for catalysis. Nature. 2014;410:780–6.
24. Walter NG, Harris DA, Pereira MJ, Rueda D. In the fluorescent spotlight: global and local conformational changes of small catalytic RNAs. Biopolymers. 2001;61:224–42.
25. Uhlenbeck OC. A small catalytic oligoribonucleotide. Nature. 1987;328:596–600.
26. Martick M, Lee TS, York DM, et al. Ribozyme Catal Chem Biol. 2008;15:332–42.
27. Martick M, Scott WG. Tertiary Contacts Distant from the Active Site Prime a Ribozyme for Catalysis. Cell. 2006;126:309–20.
28. Winkler W, Nahvi A, Roth A, et al. Control of gene expression by a natural metabolite-responsive ribozyme. Nature. 2004;428:281–6.
29. Arati R, Winker WC. Metabolite-binding ribozymes. Biochim Biophys Acta. 2014;1839:989–94.
30. Bingaman J, Zhang S, Stevens D, et al. The GlcN6P cofactor plays multiple catalytic roles in the glmS ribozyme. Nat Chem Biol. 2017;13:439–45.
31. Klein DJ, Wilkinson SR, Been MD, Ferré-D'Amaré AR. Requirement of helix P2.2 and nucleotide G1 for positioning the cleavage site and cofactor of the glmS ribozyme. J Mol Biol. 2007;373:178–89.
32. Reardon S. A twist on gene editing. Nature. 2020;578:24–7.
33. Abudayyeh OO, Gootenberg JS, Franklin B, et al. A cytosine deaminase for programmable single-base RNA editing. Science. 2019;365:382–6.
34. Merkle T, Merz S, Reautschnig P, Blaha A, et al. Precise RNA editing by recruiting endogenous ADARs with antisense oligonucleotides. Nat Biotechnol. 2019;37:133–8.
35. Guo P. The emerging field of RNA nanotechnology. Nat Nanotechnol. 2010;5:833–8842.
36. Link KH, Breaker RR. Aptamers and riboswitches: perspectives in biotechnology. Gene Ther. 2009;16:1189–201.

The Spliceosome: A Large Catalytic RNA

4

Cole Josefchak and Neena Grover

Contents

Keywords

Catalytic RNA · Ribozymes · RNA cleavage · RNA ligation · Magnesium-RNA

What You Will Learn
In eukarya, RNA is transcribed from a DNA template in the nucleus. The heteronuclear RNA (hnRNA) undergoes further processing, often while being transcribed, to be converted into a messenger RNA (mRNA). The spliceosome act on hnRNA to remove intervening regions (introns) and join expression regions (exons) using two transesterification reactions. The spliceosome is a large RNA–protein

(continued)

C. Josefchak · N. Grover (✉)
Department of Chemistry and Biochemistry, Colorado College, Colorado Springs, CO, USA
e-mail: c_josefchak@coloradocollege.edu; ngrover@ColoradoCollege.edu

© Springer Nature Switzerland AG 2022
N. Grover (ed.), *Fundamentals of RNA Structure and Function*, Learning Materials in Biosciences, https://doi.org/10.1007/978-3-030-90214-8_4

complex composed with over 200 different proteins and five small nuclear RNA (snRNA) that assembles anew on each hnRNA. In this chapter, we will discuss U2-dependent spliceosomal assembly and splicing. We will examine the many steps involved in constitutive splicing using the available cryo-EM structures and associated modeling. The spliceosome is a ribozyme with the U6 snRNA and magnesium ions participating in the catalysis reaction.

Learning Objectives

After reading this chapter students should be able to:

- Explain the role of splicing in creating a diverse set of proteins using a limited genomic repertoire.
- Discuss the key biochemical reactions of transesterification, including the roles played by snRNP complexes.
- Create a flowchart of the key steps in spliceosome assembly and catalysis during constitutive splicing.

4.1   Introduction

In early 1960s, scientists were puzzled by the fact that the radiolabel incorporated into nuclear RNA (hnRNA) was significantly diminished by the time it reached the cytoplasm [1]. The length of the RNA was also observed to decrease as it reached the cytoplasm, with no changes observed to the ends of RNA. The electron microscopy (EM) skills of several women scientists played a role in solving these mysteries [2–7]. When a DNA template was hybridized to its corresponding cytoplasmic RNA, it showed long unpaired R-shaped loops formed by the intervening sequences (introns). The introns were being removed (spliced) from the middle of hnRNA to make the cytoplasmic RNA; surprisingly, many different cytoplasmic RNA were being produced from the same hnRNA [2–7]. Thus began the discovery of RNA splicing and alternate splicing in eukaryotes. Splicing is a sort of combinatorial process of choosing segments of RNA to remove (introns) or keep (exons) from hnRNA to make the many different messenger RNA that are transported to the cytoplasm. Constitutive splicing occurs often and follows certain rules that are discussed in this chapter. The process of alternative splicing occurs when certain exons are "skipped," certain introns are retained, or an alternative donor or acceptor sites are used for the splicing reaction. There are splicing enhancer and silencer proteins that regulate splicing (trans-acting factors). The structures within the RNA also play an important role in splice site selection (cis-acting regulatory sites). It is a complex process that responds to the cellular, developmental, and environmental needs.

The ability of RNA to rearrange introns and exons led to a new appreciation for the modularity of encoded information. The exons often correspond to functional domains in proteins. These are being "shuffled" as a means of domains sharing among proteins and as a path to evolving new protein functions [7].

The process of splicing (and alternate splicing) occur in most eukaryotes and some bacteria. The spliceosome is a multi-megadalton complex with ~200 proteins and five RNA that assemble anew on each hnRNA to process it into a particular mature mRNA. RNA undergoes two transesterification steps for the cleavage and removal of intron, and the ligation of exons (Fig. 4.1).

4.2   The Group I Introns

In early 1980s, studies in a ciliated protozoan, *Tetrahymena thermophilia*, showed a 414-nucleotide intron removed itself from a 6.4 kb precursor RNA. The reaction required a free guanine (ωG) and magnesium ions. This work established the catalytic nature of RNA and the term ribozyme was coined [8]. The exogeneous guanine participates in the first reaction in a manner similar to the branch point adenine shown in Fig. 4.1. Since this guanine is not connected to the RNA, no lariat structure forms; a straight chain intron is removed in this case.

More than 1500 group 1 introns were found in bacteria (and some in eukaryotes) that share similar mechanism. Group 1 introns with open reading frames can function as mobile genetic elements that move within and between genomes. The structure of group I introns requires a pocket for binding the exogenous guanine and magnesium ions for the self-splicing reaction to occur. The crystal structure of group I intron was one of the early triumphs of RNA structural studies [9]. For decades, group I introns have served as a model system for studying RNA, in particular RNA catalysis and folding.

A new cryo-EM structure of the full length *Tetrahymena* ribozyme was recently solved in both apo and bound form (Fig. 4.2 and 4.3) [10]. The Figure 4.2c shows the secondary structure with details of base pairing interactions of the group I intron. Notice the compact, globular structure formed by this ~400 nucleotide RNA.

Another class of self-splicing introns, called the Group II introns, are found in mitochondria of yeast, fungi, and chloroplasts of unicellular organisms, such as *Chlamydomonas*. The group II introns are evolutionarily related to the spliceosome.

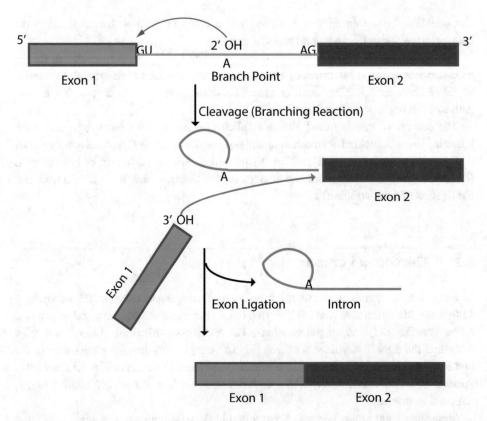

Fig. 4.1 Two Transesterification Steps of Splicing. After transcription of RNA from DNA, segments of hnRNA are removed to make a mature mRNA. There are two transesterification steps in the process. The first step involves the 2'-OH of the branch site (BS) adenine acting as a nucleophile on the 3' end of the first exon. This forms a lariat intermediate and breaks the phosphodiester linkage between the intron and the first exon (the branching reaction). The free 3'-OH of the first exon (exon 1) now serves as a nucleophile for the second reaction. It acts on the 5' end of the second exon (exon 2) to form a new phosphodiester linkage between exon 1 and 2, simultaneously freeing the 3' end of the intron (exon ligation reaction)

4.3   Splicing

Five uridine-rich small nuclear RNA: U1, U2, U4, U5, and U6 and seven associated protein, 12–35 kDa in size together form small ribonucleoproteins (snRNPs, read as "snurps"). The snRNPs are involved in pre-mRNA splicing. The in vitro splicing assays requires ATP and magnesium ions.

Most eukaryotic RNA undergo splicing. Most genes produce at least a few different messenger RNA. The frequency of splicing and alternate splicing has been reported to be 34–40% for all genes with 75–90% for multi-exon genes [11, 12]. The expression pattern

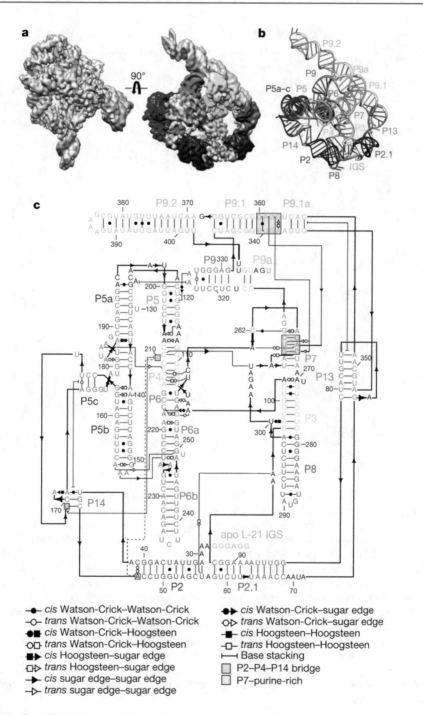

Fig. 4.2 Cryo-EM of the Apo-Form of Group I Intron. Cryo-EM structures of L-21 ScaI ribozyme (**a, b**) with the crystal structures of domains mapped into the cryo-map (**b**). A detailed secondary structure is shown (**c**). The figure is from reference [10]

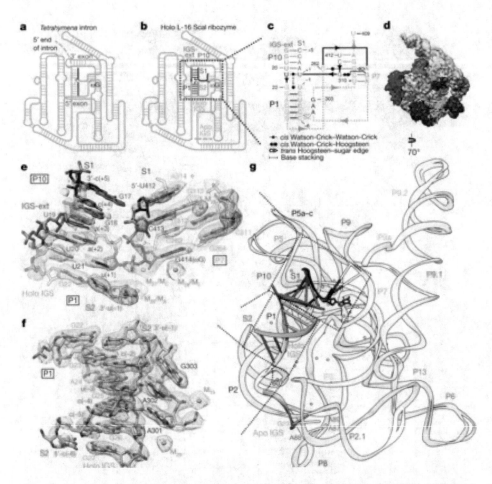

Fig. 4.3 Cryo-EM of the Holo-Form of Group I Ribozyme. The structure of the group I intron is mostly preformed in the apo-form—compare (**a**) and (**b**). The binding to two RNA substrates, mimicking the ligation reaction of the exons (holo-enzyme), show rearrangements only in the core domain. The secondary structures of the intron with bound substrates S1 (with attached terminal guanine) and S2 (the cleaved 5′ exon with a thiophosphate substitution at the active site). The 5′ end of S1 mimics intron's 3′ terminal; 3′ end of S1 mimics 3′ exon and forms P10; S2 mimics the 5′ exon and forms P1. The internal guide sequence (IGS) (dark pink or brown regions in g) binds to the two substrates (dark purple or green) to form P10, 4 bp helix (**c, e**), and P1, 6-nt helix (**c, f**). The 2-bp P7 places the guanine in the guanine binding site (the exogeneous guanine binding site). The contoured cryo-EM is shown in (**d**). The superposition of cryo-EM images of apo and holo forms (**g**) show the formation of pseudoknot structures in the catalytic core with the IGS. The IGS helps to bring all the key components (exon 1, exon 2, and magnesium ions) in close proximity (deep pink and sandy brown) for catalysis (the apo-form IGS is shown in gold in (g)) The guanine binding site (guanine shown as black sticks), on the 5′-end of S1, also undergoes conformational changes. The figure is from reference [10]

of genes has a strong impact on essential regulatory processes such as, signal transduction or chromatin remodeling. Splicing allows increasing protein diversity at a minimal cost to the organism. It is argued that alternate splicing of RNA was key to the development of complex organisms.

A few genes undergo many different splicing events. The *Dscam* gene (Down syndrome cell adhesion molecule) from *Drosophila* contains 95 alternative exons to produce trans-membrane proteins involved in cell adhesion in the nervous system. These proteins are expressed differentially during development, with greater expression during the fetal brain development to guide the new nerves to bind to the correct targets. *Dscam* gene is known for producing as many as 38,016 different isoforms of mRNA, the most known for any gene [13]. The *TTN* gene for giant muscle protein, Titin, has 363 exons and is important for heart functions. Titin proteins can be longer or shorter depending on skeletal muscle or heart related functions. Defects in splicing cause cardiac problems in humans [14]. The importance of splicing to human health was apparent from the early days of its discovery. Genome-wide analysis are showing splicing is even more wide spread and variable than previously thought.

4.4   The Spliceosome

The spliceosome is a multi-megadalton ribonucleoprotein (RNP) complex which consists of five small nuclear ribonucleoproteins (snRNPs, read as "snurp"), a nineteen complex (NTC), NTC-related proteins (NTR), and other proteins. RNP and snRNP complexes are characterized by their composition of RNA and RNA-binding proteins [15–24]. The five snRNP complexes contain a strand of a small nuclear RNA (snRNA) with sequences that are rich in uracil nucleotides and are named accordingly: U1, U2, U4, U5, and U6. Each snRNP contains an snRNA, seven Sm or LSm proteins, and a number of other proteins specific to each. The snRNAs (except U6) contain a trimethylated guanine nucleotide (m^3G) at the 5' end. A methylated adenine (m^6A) is present in U2, U4, and U6 snRNAs.

The conversion between the different spliceosomal complexes requires eight conserved RNA-dependent ATPase/helicases: Prp5, UAP56, Prp28, Brr2, and several DEAH-box ATPase/helicases that are involved in pre-mRNA processing (Prp 1, Prp16, Prp22) and Prp 43. Prp 16 and Prp 22 are also required for alternate branch-point sequence (BPS) selection and 3'-splice site (3'SS) selection.

The conformation of the spliceosome is dynamic and its structure changes throughout the assembly and splicing process. The spliceosome is not a singular complex. The proper function of the spliceosome relies on the snRNP complexes and proteins assembling on or dissociating from the pre-mRNA strand in a specific order. Although the reactions of splicing are relatively straightforward, the assembly of all RNA, proteins, and associated rearrangements are complex and involve a multitude of interactions. The interactions within the spliceosome stabilize the complex, position the pre-mRNA strand, and link various domains. We will only discuss the canonical pathway within constitutive splicing by the major spliceosome.

Recognition Sites on RNA A pre-mRNA strand contains alternating exon and intron sequences that are acted upon by the spliceosome. The phosphodiester bond connecting the 5' end of the intron to the 3' end of the 5' exon is known as the 5'-splice site (5'SS), with a sequence requirement of GU at this site. U1 binds to 5'SS using a complementary sequence. On the 3' end of the intron, the phosphodiester bond connects the intron to the 5' end of the 3' exon, the 3'-splice site (3'SS), with an AG sequence requirement at this site. Within the intron is the branch point sequence (BPS) which is located upstream from the 3' splice site and is characterized by an adenosine nucleotide. The branch point sequence serves as the recognition site for U2 snRNA. The branch site is followed by polypyrimidine tract (PPT). The intron's BPS and 3'-SS regions are defined early in the splicing cycle [18]. The branch site sequence has the consensus: YUNAY (Y = pyrimidines and N = any nucleotide). This sequence sometimes is present multiple times; the copy that is utilized depends on the cell type. In yeast, cooperative recognition of BPS by heterodimer of splicing factor 1 (SF1) and PPT by U2 associated factor (U2AF) occurs. The sequences and structures of RNA that surround the splice sites and proteins that bind to these (cis and trans factors) all play a role in site selection and assembly of the spliceosome. As expected, the story is a more intricate and involved for site selection between splicing and alternate splicing. As genome-wide analyses are performed, alternate sites splicing are being discovered. These likely exist to enhance fidelity and regulation of splicing. The general principles presented here will orient us to the complexity of the processes involved in RNA splicing.

Overview of Spliceosomal Assembly The spliceosomal assembly begins with recognition of 5'SS, BPS, and 3'SS. The U1 snRNP and other splicing factors associate with the pre-mRNA strand to assemble the E complex. The U2 snRNP interacts with both the pre-mRNA strand and the U1 snRNP to make the A-complex. The recruitment and assembly of the U4/U6/U5 tri-snRNP complex follows the U2 snRNP interaction, forming the B complex. The B complex undergoes further conformational changes to become activated, B^{act} to B* that causes unpairing of U4 and U6 along with release of U1 and U4 as other proteins join the complex. The U6 has the catalytic magnesium ion bound that is involved in the first transesterification reaction producing exon 1 and lariat-intron-exon 2 intermediates. The spliceosome remodels to complex C allowing U5 snRNA to align exons 1 and 2 for ligation. The C complex requires activation in order to catalyze the second transesterification reaction. The ligation results in an mRNA strand that is ready for translation, and an intron lariat structure that is still connected to the spliceosome. The mRNA strand is released from the spliceosome upon completion of the ligation reaction and the spliceosome complex then dissociates, releasing the intron lariat. A detailed overview of the splicing process is shown in Fig. 4.4 [25].

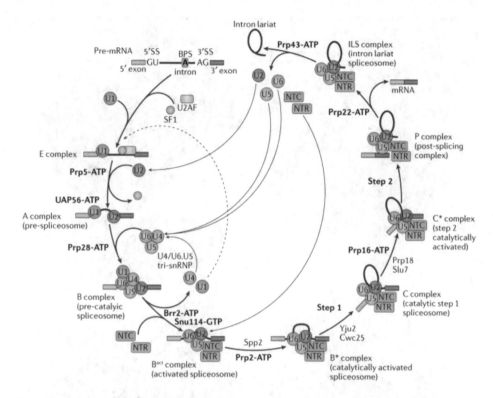

Fig. 4.4 Spliceosomal Assembly. A detailed view of the steps in spliceosomal assembly. The process begins with a single pre-mRNA and results in excision of intron and joining of the exons. The 5'-splice site (5'-SS), the 3'-splice site (3'-SS), and the branch point adenosine (BPS) are shown on the pre-mRNA. Various snRNPs and protein that join or leave the spliceosome during different steps are shown. Figure is from reference [25]

The E Complex The 5'SS often has GU nucleotides that are recognized by a complementary sequence on U1 snRNA. As U1 snRNP associates with the 5' splice site, two splicing factors, SF1 and U2AF, recognize and interact with the 3'splice site and the branch point, respectively. This is the first step of spliceosome assembly, the formation of the E complex. The base pairing interactions in the first step are spontaneous. The displacement of SF1 enables recruitment of the U2snRNP and requires ATP hydrolysis.

The Formation of A Complex The U2 snRNP selects the 3' of the splice site, binds the BPS adenine, and the PPT [15–26]. The U2 snRNA base pairs with BPS in an ATP-dependent manner to yield a pre-spliceosomal A complex. The invariant adenine of the branch site is unstacked from the U2/BPS helix and will later serve as a nucleophile during splicing reaction.

The structure of the U2 snRNP has two lobes (Fig. 4.5) connected by a bridge, a smaller 3'-module and larger 5'-module. The 5'-module contains the splicing factor 3B, SF3B

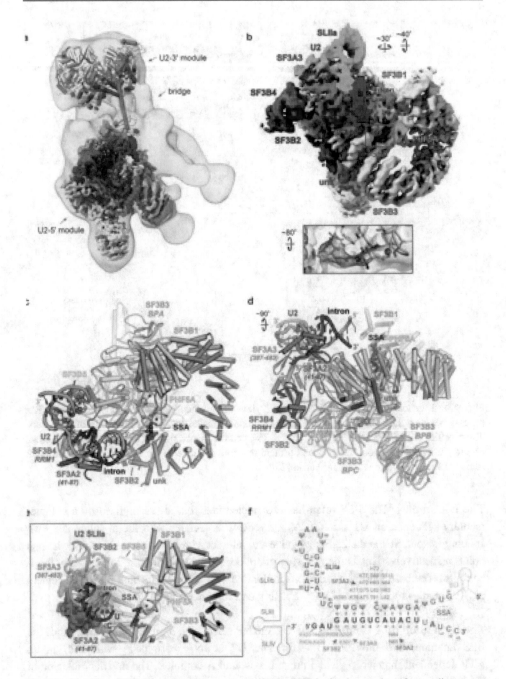

Fig. 4.5 The U2 snRNP. The cryo-EM structures of U2snRNP in the A-complex of pre-spliceosome with U2-snRNA and U2-snRNP structures superimposed (**a–e**) and the secondary structure of U2/intron (light blue/dark blue) with interacting residues marked (**f**). Figure is from reference [26]

whose core subunit is called SF3B1^{HEAT} and has a HEAT domain that binds the branch point helix, trapping the branch point adenine in a hinge pocket. This domain has an open and closed conformations. The closed conformation is seen in the spliceosome where it stabilizes the U2/intron helix and the PPT tract, simultaneously. The branch point stem loop has to unwind and interact with the intron to form a U2/intron duplex, which remains unchanged throughout A to B^{act} complex The duplex is confined to 16 base pairs by two splicing factors, SF3A and SF3B. This step helps in selection of the intron by a toe-hold mediated strand invasion.

The Formation of the U4/U6/U5 Tri-snNP   The tri-snRNP interacts with mRNA bound to U1 and U2 to make a catalytically active spliceosome (Fig. 4.6) [27]. The tri-snRNP complex is stabilized by protein–protein, RNA–RNA, and RNA–protein interactions. The

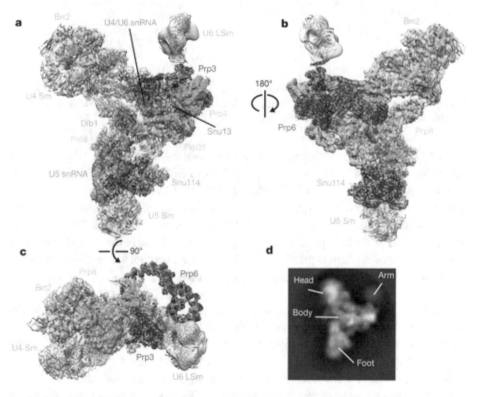

Fig. 4.6 Cryo-EM of the Tri-snRNP. The U4, U5, and U6 snRNP form a U4/U6/U5 tri-snRNP prior to interacting with the spliceosome This 1.5 megadalton spliceosomal complex is an assembly of nearly 30 proteins, including three key proteins that are important for activation of the spliceosome: Prp8, Brr2, and Snu114. Prp8 is important for substrate positioning; Brr2 is a helicase that unwinds the duplex between U4/U6 snRNA and is regulated by GTPase Snu114. The tri-snRNP components are color coded and labeled and shown in three different rotations (**a–c**); it forms a triangular structure with head, body, arm and foot regions (**d**). Figure adapted from [27]

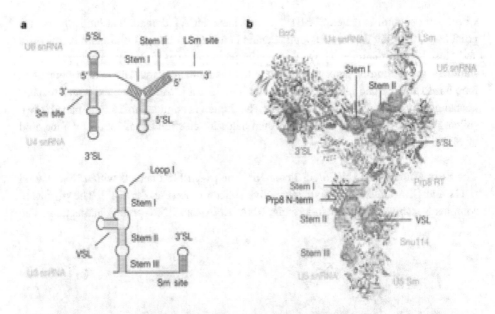

Fig. 4.7 The U4/U6.U5 Tri-snRNP. (**a**) The secondary structures of the U4/U6 and U5 snRNAs. (**b**) The cryo-EM structure of the tri-snRNP complex. The regions of RNA are color coded in (**a**) and (**b**). Variable stem loop (VSL) and stem loop (SL). Figure is from [27]

U4 and U6 snRNA are extensively base paired to create dsRNA helix (Fig. 4.7). The single-stranded region of U4 snRNA is loaded on to Brr2 helicase active site, ready to unwide its stem-loop (labeled as Sm site in Figure 4.7a). All the snRNA components in the spliceosome have secondary structural features that serve as recognition sites for proteins. The U4 snRNA contains a 3′ stem loop (3′ SL) that interacts with Brr2. The interaction correctly positions the Brr2 protein which plays an essential role in the activation of the B complex. Without the correct positioning of Brr2, the B complex will not be activated to B^{act}, thus preventing the first catalytic reaction. Figure is from [27].

Unlike the RNA–RNA interaction between U4 and U6 snRNP, the U5 snRNP is only connected to the complex by interactions with Prp8, one of the U5 snRNP proteins. Prp8 is positioned at the center of the tri-snRNP complex and functions to hold the three snRNP complexes together. The Prp8 also stabilizes stem I, stem II, and the variable stem loop (VSL) of U5 snRNA; it positions its loop1, which aligns the exons for splicing.

During spliceosome assembly and activation, the Prp8 domain plays a key role in the moving the RNA into the active site. The ACAGAGA sequence of U6 snRNA, the U6 snRNA region that base pairs with U2 snRNA, and the U5 snRNA loop I are all found near the active site during the catalytic reactions and are necessary for the splicing process.

The Formation of the B Complex The formation of the B complex is the next step of spliceosome assembly and it involves the recruitment of both the U2 snRNP and the U4/U6/U5 tri-RNP complex [15–28]. The U2 snRNP is recruited first and it associates with the E complex in two ways. The U2 snRNA binds with the pre-mRNA at the branch point sequence through complementary base pairing. The U2 snRNP protein subunits also interact with the outer surface proteins of the U1 snRNP. Shortly after the U2 snRNP binds to the complex, the U4/U6/U5 tri-snRNP complex is recruited and, once assembled, the spliceosome structure is known as the B complex. The assembly of the B complex requires ATP hydrolysis unlike the assembly of the E complex.

Upon the formation of the B complex, one of the tri-snRNP proteins, Prp28, disrupts the base pairing interaction between the U1 snRNA and the 5' splice site. This disruption destabilizes the U1 snRNP complex, causing it to dissociate from the spliceosome. The 5' splice site then interacts with the ACAGAGA region of the U6 snRNA. Without the U1 snRNP, the conformation of the remaining snRNP complex is concave, with the U2 snRNP protruding over the top of the tri-RNP complex (Fig. 4.8). The 5' domain of the U2 snRNP interacts with the tri-snRNP complex through two interfaces. In interface 1, the interactions consist of a splicing factor subunit, SF3b, binding to both the U4/U6 snRNP protein, Prp3, and the U2/U6 helix II. Interface 2 involves a couple of weak protein interactions between the N- and C-terminal of Brr2 as well as the N-terminal of Brr2 and surrounding protein residues. The weak interactions at interface 2 allows for more flexibility and contributes to the dynamic nature of the spliceosome. The U2 snRNP structure is bipartite and the two domains of U2 are bridged by a splicing factor, SF3a. Together, SF3a and SF3b form a network of protein-RNA interactions that assist in escorting the U2 snRNA and the intron, along with many other stabilizing interactions occurring in the B complex. Prior to activation of the B complex, a network of intricate protein interactions are stabilizing both the helicase Brr2 protein on U4 snRNA, and base paring interaction between the 5' splice site and the U6 snRNA.

Spliceosome Activation, Complex B^{cat} to B* Conformational changes must occur in the B complex to correctly position the various components for the splicing reaction to proceed [15–29]. This process is known as activation. Activation of the B complex causes the U1 snRNP and the U4 snRNP to dissociate and the resulting B complex contains U2, U5, and U6 snRNPs. The U1 snRNP dissociates upon the tri-snRNP complex binding as the U1 snRNA-5' splice site interaction is disrupted by Prp28. The free 5' splice site interacts with the ACAGAGA region of the U6 snRNA prior to the first catalytic reaction. The U6 snRNA-5' splice site interactions serves as a checkpoint to ensure the B complex was assembled properly, prior to the unwinding of the U4/U6 snRNA double helix. The dissociation of the U4 snRNP begins with the unwinding of the U4/U6 snRNA double helix, which is catalyzed by the helicase protein subunit of the U5 snRNP, Brr2. Prior to unwinding the RNA helix, Brr2 is preloaded onto the single-stranded region of the U4 snRNA located downstream of stem I of U4/U6 snRNA. The disruption of the U4/U6

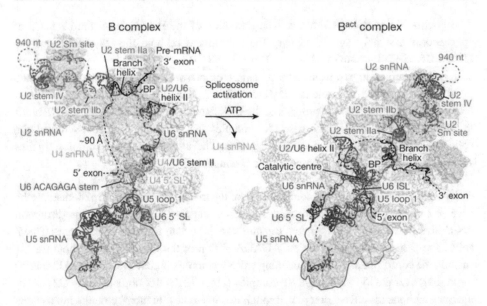

Fig. 4.8 Cryo-EM Structures of B and B^act Complexes. A structural comparison of the B (left) and B^act complexes (right). The snRNA strands are superimposed onto the cryo-EM map. The snRNA strands are shown in the same color as their corresponding snRNP complex; the pre-mRNA strand is shown in black. Follow the movement of the U6 between B and B. Figure is from reference [28]

snRNA base pairing is ATP-dependent and the unwinding can occur in the absence of pre-mRNA thus this happens independently of pre-mRNA splicing.

A structural comparison of the B and B^act complex is shown in Fig. 4.8. Upon the activation of the B complex, the U6 snRNA tethers the U2 and U5 snRNPs. In the B^act complex, without the U4 snRNP complex, the U6 snRNA forms a double-stranded RNA helix with U2 snRNA. The initial base paring interactions between the U2 snRNA and the intron at the branch point that occurred during assembly of the B complex are still present upon activation. At the catalytic core of the B^act complex, only the key components are present—the ACAGA sequence on U6 snRNA interacting with the 5′ splice site, and loop I of the U5 snRNA base paired with the 3′ end of the 5′ exon. Figure 4.9 shows a schematic drawing of interactions occurring between RNA in the B^act complex. The interaction between the U5 snRNA and the 5′ exon holds the 5′ exon at the active site.

Other RNA–protein interactions also stabilize the complex. The components not involved in the catalytic reactions are located away from the active site, for example, the 5′ and 3′ ends of the U2, U5, and U6 snRNAs. The U6 internal stem-loop (ISL) region binds to magnesium ions. Metal ions play a critical role in stabilizing correct RNA structures and are essential for the two transesterification reactions, which will be discussed later.

The B^act complex is converted to B* by the action of ATPase/helicase Prp2 complex; Prp2 interactions allows the first transesterification reaction to occur at the 5′-splice site,

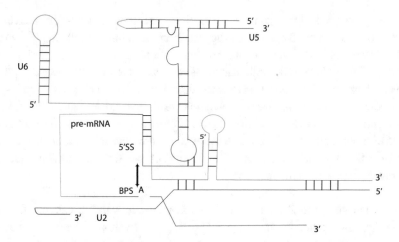

Fig. 4.9 RNA Base Pairing in B^{act} Complex. A schematic depiction of RNA that are base paired in the B^{act} complex. U6 is now base paired to U2. U5 is stabilizing the 5′ exon. The branch point and splice site are further apart at this point [29]. The Figure is based on [29]

yielding the free 5′-exon and the intron lariat structure. A adenosine nucleotide in the branch point in pre-mRNA is bulged out from the helix and forms a hydrogen bond with the uracil nucleotide located two nucleotides downstream. This interaction allows the 2′-hydroxyl group of the adenosine to protrude toward the 5′-end of the intron. The first catalytic reaction proceeds due to the close proximity between the adenosine's 2′-hydroxyl group and the 5′ splice site, resulting in a lariat structure for the intron with a 2′-5′ phosphodiester bond. A model for proposed changes in the spliceosome are shown in Fig. 4.10.

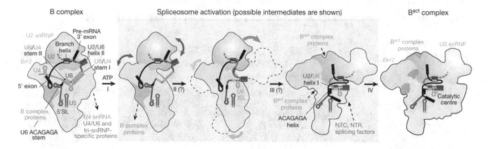

Fig. 4.10 A Model for B^{act} Complex Formation. The helicase activity of Brr2 results in release of U4 snRNA (yellow) along with U4/U6 and tri-snRNP-specific proteins (I). This is followed by U6 ISL formation (pink portion of U6) and ACAGAGA stem unfolding (II). Next 5′-exon (black box) and U5 loop (blue) I base pairing (III), followed by the binding of NTC, NTR and B^{act} complex proteins (IV) to form the B^{act} complex. Proposed intermediates are drawn with gray shaded background. Figure is from [28]

B^{act} to C Complex Upon completion of the first transesterification reaction, the spliceosome is now in the C complex, which has structural similarities to the B^{act} complex. Loop I of U5 snRNA continues to hold the 5'-exon in place through base pairing, and the ACAGA region of U6 snRNA continues to interacts with the intron nucleotides [30]. The NineTeen complex (NTC) and NTC-related complex (NTR), along with other splicing factors, work together to stabilize the catalytic core in both the B^{act} and C complex. The free 5'-exon is located in a narrow channel that is formed upon spliceosome activation and is in-between two domains of Prp8. The metal ions within the spliceosome are rearranged by the U6 snRNA during activation. The U6 snRNA positions the catalytic metal ions appropriately for the catalytic reactions.

The C^{act} Complex The C complex must be activated in order to catalyze the second transesterification reaction by the action of Prp16. The activation of the C complex causes conformational changes that bring the necessary splicing components into the active site.

The active site of the C^{act} complex brings the 3' OH of the 5' exon toward the 3' splice site with the assistance of two helicase proteins, Prp16 and Prp22. In the C complex, Prp16 is located downstream of the branch point (Fig. 4.11a). During activation, Prp16 rearranges causing destabilization of two proteins that are bound at the active site. During this destabilization, the 3' splice site enters the active site and is stabilized by interacting with two other proteins, Prp18 and Slu7. In addition, Prp22 interacts with the 3' exon, forming the C^{act} complex. With the help of the appropriately positioned metal ions, the close positioning of the 3' OH to the 3' splice site allows the second transesterification reaction to take place, yielding two products, the intron lariat and the mRNA strand. An illustration of the activation of the C complex is shown in (Fig. 4.11b). The B^{act} and the C^{act} complex contain the key conformational elements required to catalyze the two reactions at the active site. With all the necessary components in close proximity, the reactions can be catalyzed efficiently. The protein–protein and protein–RNA interactions occurring away from the catalytic core function to stabilize the entire complex, including the RNA core.

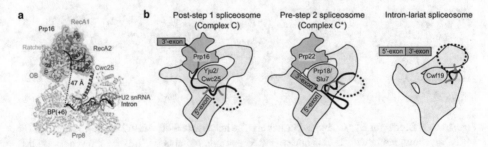

Fig. 4.11 The Role of Helicases in the Active Site. (**a**) The position of the intron sequence that is downstream from the branch site. The 3' end of the intron exits by a channel in Prp8 and protrudes toward Prp16. (**b**) The step-by-step illustration of the C complex activation and catalysis for the second reaction. Figure from [30]

Disassembly of the Spliceosome After the second transesterification reaction occurs, Prp22 catalyzes the dissociation of the newly synthesized mRNA strand from the spliceosome. Prp22 utilizes the energy from ATP hydrolysis to disrupt the residual interactions of the mRNA strand. The post-catalytic spliceosome complex contains the intron lariat, U2, U5, and U6 snRNP. Shortly after the mRNA strand is released, the post-catalytic spliceosome complex dissociates into the respective snRNP complexes and the intron lariat structure. The intron lariat is then tagged for degradation. The snRNP complexes are reused in another splicing reaction.

4.5   The Spliceosome Is a Ribozyme

Spliceosome, like the ribosome, is a large RNA–protein complex. Since the discovery of the Group I and II introns, the role of RNA in spliceosome catalysis was suspected. The assembly and catalysis of spliceosome require dynamic rearrangements of many different factors as discussed above. It took until 2013 to convincingly show that RNA indeed performs the catalysis reactions, making it a large ribozyme [31].

The spliceosome catalyzes two transesterification reactions. The mechanisms for both reactions involve a nucleophile attacking an electrophile, creating a new bond at the expense of a leaving group (Fig. 4.12) [31]. In the first reaction, the $2'$ hydroxyl group of the adenosine at the branch point acts a nucleophile and attacks the phosphodiester bond at the $5'$ splice site. The leaving group is the $3'$ oxygen at the end of the $5'$ exon. The first catalytic reaction yields: A) an the intron lariat created by a $2'$-$5'$ phosphodiester bond between the branch point adenosine and the $5'$ end of the intron, and B) the $5'$ exon is separated from the pre-mRNA strand leaving the $3'$ hydroxyl group exposed. In the second

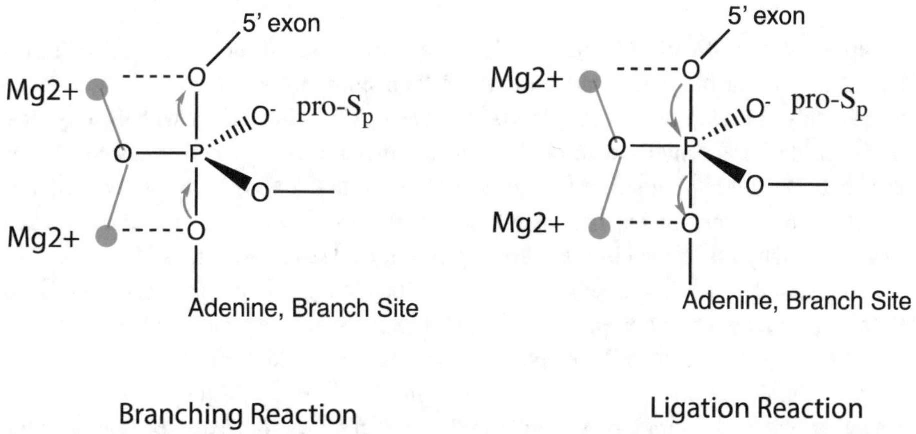

Branching Reaction   Ligation Reaction

Fig. 4.12 The Transesterification Reactions in the Spliceosome. The branching and ligation reactions require formation of a trigonal planar transition state that is stabilized by two magnesium ions. Figure based on [31]

transesterification reaction, the 3′ hydroxyl group of the 5′ exon acts as the nucleophile and attacks the phosphodiester bond at the 3′ splice site to ligate the exons. The leaving group in this reaction is the oxygen at the 3′ end of the intron. The second reaction yields a separated intron lariat structure and an mRNA strand containing the two exon sequences.

The two transesterification reactions occur at the same active site in the spliceosome. Magnesium ions are required in the active site for catalysis, and they function to stabilize the transition states. The positive charge on the magnesium ion interacts with the negatively charged oxygen leaving group. This interaction reduces oxygen's attraction to the positively charged phosphate, which allows it to dissociate more readily. In both the catalytic reactions, magnesium ions interact with nucleotides in U6 snRNA and the pre-mRNA strand. The snRNA and pre-mRNA interact with the catalytic metal ion through the oxygen attached to the phosphate backbone. There are two non-bridging oxygens attached to each phosphate, and their respective positions are shown in Fig. 4.12. One oxygen lies in the pro-R_p position and the other oxygen in the pro-S_p position (as discussed in Chap. 1).

First Transesterification Reaction In the branching reaction, the U6 snRNA oxygens that are involved in binding a magnesium ion are in the following positions: G78 pro-S_p, U80 pro-R_p, U80 pro-S_p. One of the metal ions, M1, interacts with two oxygens in the pre-mRNA strand that are located at the 5′ splice site. The oxygen participating in the phosphodiester bond that binds the intron to the 5′ exon acts as the leaving group. The other pre-mRNA oxygen is non-bridging and is located at the pro-R_p position. Furthermore, the positioning of M1 results from interactions with the oxygens of the U6 snRNA that are located at the U80 pro-R_p and the G78 pro-S_p position. The other metal ion, M2, interacts with the pre-mRNA strand through the non-bridging oxygen at the 5′ splice site in the pro-R_p position. Analogous to M1, M2 is positioned by the U6 snRNA and specifically, the oxygen at the U80 pro-S_p position (Fig. 4.12).

Second Transesterification Reaction The second transesterification reaction, exon ligation, takes place at the same active site as the branching reaction. One difference in the second catalytic reaction is that only the second metal, M2, is interacting with the oxygens. The U6 snRNAs are involved in positioning the metal ions in a similar fashion to the branching reaction by binding to the oxygens attached to the phosphate backbone. There are only two oxygens in the U6 snRNA strand that are involved in the exon ligation reaction and they are located in the following positions: U80 pro-S_p and A59 pro-S_p. The same magnesium ion, M2, is interacting with both these oxygen molecules. The U6 snRNA oxygens positions M2 appropriately to allow M2 to also interact with the oxygen participating in the phosphodiester bond located at the 3′ splice site of the pre-mRNA strand. It should be noted that the oxygen at the U80 pro-S_p position is binding and orientating the magnesium ion, M2, in both the branching and the exon ligation reaction. The U6 snRNA oxygens involved in binding a magnesium ion during the catalytic reactions mainly function to correctly position the metal ions. The proper positioning of the metal ions allows them to sufficiently interact with the leaving group oxygens in the pre-mRNA strand. The main role of metals ions interacting with the pre-mRNA oxygens is to provide stability to the negatively charged leaving group during the nucleophilic attack.

Take Home Message

- Splicing occurs on most hnRNA in the nucleus to generate different mRNA from the same transcript.
- Each hnRNA may have multiple introns and exons. The choice of splice sites is determined by structures within RNA and by myriad proteins that bind to it.
- The assembly of the spliceosome requires several snRNP to interact with multiple ATPase/helicase. Various other proteins either enhance or silence particular splicing sites.
- The complex and intricate structures within the splicesome have been imaged crystallography and now using cryo-EM. The various interactionst have to occur between RNA-RNA and RNA-proteins to allow correct positioning of the active sites in the spliceosome. In depth understanding of the splicing steps is necessary if we are to understand and treat defects in splicing that lead to diseases.

References

1. Harris H, Watts JW. The relationship between nuclear and cytoplasmic ribonucleic acid. Proc R Soc Lond B Biol Sci. 1962;156:109–12.
2. Berget SM, Moore C, Sharp PA. Spliced segments at the 5′ terminus of adenovirus 2 late mRNA. Proc Natl Acad Sci U S A. 1977;74:3171–5.
3. Berk AJ, Sharp PA. Sizing and mapping of early adenovirus mRNAs by gel electrophoresis of S1 endonuclease-digested hybrids. Cell. 1977;12:721–32.
4. Chow LT, Gelinas RE, Broker TR, Roberts RJ. An amazing sequence arrangement at the 5′ ends of adenovirus 2 messenger RNA. Cell. 1977;12:1–8.
5. Darnell JE Jr. Implications of RNA–RNA splicing in evolution of eukaryotic cells. Science. 1978;202:1257–60.
6. Early P, Rogers J, Davis M, et al. Two mRNAs can be produced from a single immunoglobulin chain by alternative RNA processing pathways. Cell. 1980;20:313–9.
7. Berk AJ. Discovery of RNA splicing and genes in pieces. Proc Natl Acad Sci. 2016;113:801–5.
8. Bass B, Cech T. Specific interaction between the self-splicing RNA of Tetrahymena and its guanosine substrate: implications for biological catalysis by RNA. Nature. 1984;308:820–6.
9. Cate JH, Gooding AR, Podell E, et al. Crystal structure of a group I ribozyme domain: principles of RNA packing. Science. 1996;273:1678–85.
10. Su Z, Zhang K, Kappel K, et al. Cryo-EM structures of full-length Tetrahymena ribozyme at 3.1 Å resolution. Nature. 2021;596:603–7.
11. Kashyap L, Sharma RK. Alternative splicing: a paradoxical qudo in eukaryotic genomes. Bioinformation. 2007;2:155–6.
12. Rotival M, Quach H, Quintana-Murci L. Defining the genetic and evolutionary architecture of alternative splicing in response to infection. Nat Commun. 2019;10:1671.

13. Celotto AM, Graveley BR. Alternative splicing of the drosophila Dscam pre-mRNA is both temporally and spatially regulated. Genetics. 2001;159:599–608.
14. Savarese M, Jonson PH, Huovinen S, et al. The complexity of titin splicing pattern in human adult skeletal muscles. Skelet Muscle. 2018;8:11.
15. Will CL, Lührmann R. Spliceosome structure and function. Cold Spring Harb Perspect Biol. 2011;3:a003707.
16. Guthrie C. Messenger RNA splicing in yeast: clues to why the spliceosome is a ribonucleoprotein. Science. 1991;253:157–63.
17. Brow DA, Guthrie C. Spliceosomal RNA U6 is remarkably conserved from yeast to mammals. Nature. 1988;334:213–8.
18. Lee Y, Rio DC. Mechanisms and regulation of alternative pre-mRNA splicing. Annu Rev Biochem. 2015;84:291–323.
19. Bindereif A, Green MR. An ordered pathway of snRNP binding during mammalian pre-mRNA splicing complex assembly. EMBO J. 1987;6:2415–24.
20. Feltz C, Anthony K, Brilot A, Pomeranz Krummel D. Architecture of the spliceosome. Biochemistry. 2012;51:3321–33.
21. Raghunathan PL, Guthrie C. RNA unwinding in the U4/U6 snRNPs requires ATP hydrolysis and the DEIH-box splicing factor Brr2. Curr Biol. 1998;8:847–55.
22. Schwer B. A conformational rearrangement in the spliceosome sets the stage for Prp22-dependent mRNA release. Mol Cell. 2008;30:743–54.
23. Weber S, Aebi M. In vitro splicing of mRNA precursors: 5′ cleavage site can be predicted from the interaction between the 5′ splice region and the 5′ terminus of U1 snRNA. Nucleic Acids Res. 1998;16:471–86.
24. Wu J, Manley JL. Mammalian pre-mRNA branch site selection by U2 snRNP involves base pairing. Genes Dev. 1989;3:1553–61.
25. Shi Y. Mechanistic insights into precursor messenger RNA splicing by the spliceosome. Nat Rev Mol Cell Biol. 2017;18:655–70.
26. Cretu C, Gee P, Liu X, et al. Structural basis of intron selection by U2 snRNP in the presence of covalent inhibitors. Nat Commun. 2021;12:4491.
27. Nguyen TH, Galej WP, Bai XC, et al. The architecture of the spliceosomal U4/U6.U5 tri-snRNP. Nature. 2015;523:47–52.
28. Plaschka C, Lin P, Nagai K. Structure of a pre catalytic spliceosome. Nature. 2017;546:617–21.
29. Yan C, Wan R, Bai R, et al. Structure of a yeast activated spliceosome at 3.5 Å resolution. Science. 2016;353:904–11.
30. Galej W, Wilkinson M, Fica S, et al. Cryo-EM structure of the spliceosome immediately after branching. Nature. 2016;537:197–201.
31. Fica SM, Tuttle N, Novak T, et al. RNA catalyses nuclear pre-mRNA splicing. Nature. 2013;503:229–34.

A Genomics Perspective on RNA

5

Juliana C. Olliff, Jia A. Mei, Kristie M. Shirley, and Sara J. Hanson

Contents

J. C. Olliff
Department of Molecular Biology, Colorado College, Colorado Springs, CO, USA
e-mail: j_olliff@coloradocollege.edu

J. A. Mei · K. M. Shirley · S. J. Hanson (✉)
Department of Molecular Biology, Colorado College, Colorado Springs, CO, USA
e-mail: j_mei@coloradocollege.edu; k_shirley@coloradocollege.edu;
shanson@coloradocollege.edu

© Springer Nature Switzerland AG 2022
N. Grover (ed.), *Fundamentals of RNA Structure and Function*, Learning Materials in
Biosciences, https://doi.org/10.1007/978-3-030-90214-8_5

Keywords

mRNA-seq · Transcriptome · High-throughput sequencing · Genomics · Sequencing-by-synthesis · Differential expression · Long-read sequencing

What You Will Learn
- Introduction to sequencing technologies that allow for whole genome exploration of RNA, and how they have shaped the scale of RNA studies.
- Overview of the general workflow for a genome-scale experiment with RNA, including library preparation, sequencing, and analysis.
- Applications for sequencing methods to study a variety of RNA types, including messenger RNA and long or small noncoding RNA.
- Applications for sequencing methods to study processes involving RNA, including translation, transcription initiation, and RNA–protein interactions.
- Discussion of recent technologies that are expanding in the field, including long read, direct RNA sequencing, and single-cell sequencing.

5.1 From Transcript to Transcriptome: The Impact of Next-Generation Sequencing Methods on the Study of RNA

In recent decades, the development of tools for generating nucleic acid sequences at a high-throughput scale revolutionized the field of molecular biology. At the turn of the twenty-first century, methods for studying RNA were largely limited to the examination of individual genes (Fig. 5.1). The changes in expression of a single gene, for example, could be studied through Northern blots [1] and reverse transcriptase polymerase chain reaction (RT-PCR) [2]. Although powerful methods that allow for detailed investigation of a particular transcript, their reach is limited to the study of one or a few genes at a time. Scaling these studies to examine every gene in the genome is cost prohibitive and time

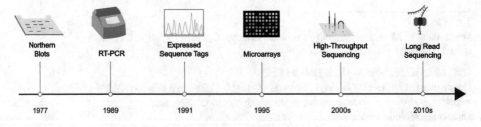

Northern Blots — RT-PCR — Expressed Sequence Tags — Microarrays — High-Throughput Sequencing — Long Read Sequencing

1977 — 1989 — 1991 — 1995 — 2000s — 2010s

Fig. 5.1 Timeline of methods for studying gene expression

consuming. The first methods for studying RNA at a larger scale included cataloguing of Expressed Sequence Tags (a type of sequencing library produced from complementary DNA, or cDNA) by Sanger sequencing methods [3] and the development of microarrays [4]. Through these methods, the first glimpses of the genomic view of RNA were gleaned, but they remained limited by their cost, speed, and the resources required for their implementation.

With the advent of high-throughput sequencing (HTS), the volume of sequencing data that could be produced began to rise exponentially as the cost of sequencing dropped [5]. This led to a dramatic expansion in the scale of RNA studies, in which the perspective can be broadened to the transcriptome, or the entirety of a cell's or tissue's RNA content. The accessibility and affordability of these methods have led to them becoming standardized in the field, and have spurred the further development of cutting-edge technologies to delve into the complexities of RNA expression and function in the cell with much higher resolution [6].

HTS includes a set of methods and technologies that allow for simultaneous sequencing of millions of short (~50–300 bp), uniformly sized nucleic acid fragments. This chapter serves as an introduction to how HTS is used to study RNA at the transcriptome level. An overview of the workflow commonly used for high-throughput sequencing of messenger RNA (mRNA-seq) is presented, as well as modifications to the procedure that allow for the study of different types of RNA and cellular processes that involve RNA. Finally, more recent innovations in sequencing technologies and their applications are presented.

5.2 Using mRNA-Seq to Investigate the Protein-Coding Transcriptome

Given its role as the intermediate between DNA and protein, messenger RNA (mRNA) has been the focus of large amount of transcriptome work. mRNA sequencing (mRNA-seq) has been used to address a wide array of biological questions. How does gene expression change in response to environmental conditions, or during different stages of development? When and where are different isoforms of an RNA transcript expressed? Which genes are regulated by a particular transcription factor?

Sequencing-by-synthesis (SBS), developed by Illumina, Inc., is a popular HTS method for performing mRNA-seq. Like other HTS methods, SBS technology allows for massively parallel sequencing of short nucleic acid molecules. Depending on the specific sequencing platform used, an SBS run can produce as much as six trillion base pairs (6 Tb) of sequence in a single run [7]. The volume of data produced by SBS allows for accurate and robust quantification of gene expression across the transcriptome [8], which can be used for comparison of gene expression levels between samples. SBS can also allow for paired-end sequencing, in which both ends of a DNA molecule are sequenced. The additional sequencing information can be used to assemble transcriptomes de novo, or without the need for a preexisting genome sequence for an organism to be used as a reference [9]. In addition, SBS protocols can be modified to track the strandedness, or directionality, of a transcript, which distinguishes between sense and antisense transcription [10]. Careful planning of a sequencing experiment includes considering how the final dataset will be analyzed in order to prepare a sequencing library and sequencing reaction that will sufficiently address the biological question of interest [11].

This section delves into the workflow and methods commonly used to perform mRNA-seq with SBS (Fig. 5.2). RNA is first isolated from a sample and mRNA is used to generate a cDNA library. Following sequencing of the cDNA library, bioinformatics methods are used to assemble the short sequences into transcripts, and further analysis is performed, such as identifying transcripts that are differentially expressed between experimental conditions.

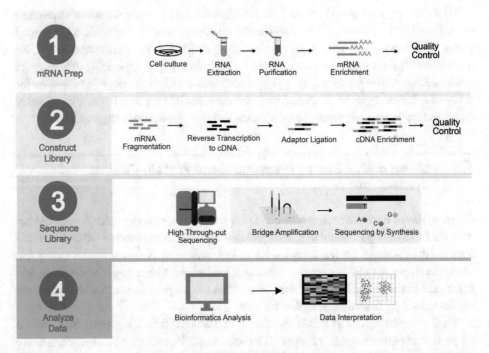

Fig. 5.2 Overview of mRNA-seq workflow. An mRNA-seq experiment using SBS technology occurs in a series of four stages. (1) mRNA is isolated from a sample or samples of interest. (2) A cDNA library is constructed from the input mRNA. Library construction includes addition of adaptor sequences that are required for the downstream sequencing reaction. (3) The cDNA library is amplified on a solid substrate and SBS determines the sequence of each fragment in the library in parallel. (4) The large amount of sequencing data is processed and analyzed to address biological questions

5.2.1   Preparing RNA for Sequencing: Isolation, Fragmentation, and Enrichment

The first step in performing an mRNA-seq workflow is isolation of RNA from a source of interest, such as a multicellular tissue or a population of unicellular organisms. This step must be performed with biological replication by preparing multiple samples that have received identical treatment (typically a minimum of three for each sample type in the experiment) to allow for statistical evaluation during data analysis. To isolate RNA, cells must be chemically lysed using detergents to disrupt cell membranes. In the case of cells with cell walls, such as plants and yeasts, mechanical disruption is also required using glass beads or mortars.

Following lysis, RNA must be isolated from other macromolecules found in the cell, including DNA, proteins, and lipids. The differing solubilities of these macromolecules are used as the basis for extraction. Proteins and lipids are removed first due to their solubility in organic solvents (phenol and chloroform, respectively). In contrast, the nucleic acids RNA and DNA are soluble in aqueous solutions, allowing for their separation from the organic solutes. When isolating RNA from a sample, acid phenol chloroform is used. The reduced pH in this method encourages the solubility of DNA in the organic phase, thereby enriching the aqueous phase specifically with RNA [12]. To further purify the RNA from the nucleic acid mixture, samples are treated with deoxyribonuclease I (DNase I), an enzyme that breaks down double-stranded and single-stranded DNA molecules without sequence specificity [13].

Working with RNA requires a great deal of care to prevent the degradation of the RNA molecules. Unlike DNA, RNA contains a hydroxyl group on the $2'$ carbon of each nucleotide that is susceptible to hydrolysis reactions that will break down the backbone of an RNA molecule. In addition, ribonucleases (RNases), enzymes that target RNA molecules for degradation, are ubiquitous in intracellular and extracellular environments. Using RNase inhibitors can help mitigate the degradation of samples, in addition to using careful lab practices and sterile technique.

The RNA purity, quality, and quantity are rigorously assessed prior to constructing a sequencing library. RNA sample purity is measured to determine the amount of genomic DNA contamination that is present in the sample following DNase treatment. The RNA will be converted to cDNA during library construction, and the presence of genomic DNA may result in incorporation of genome sequences into your library that do not reflect levels of gene expression. Low quantities or poor-quality RNA can also impact cDNA library prep. Library preparation requires established minimum amounts of RNA, and samples with quantities less than the minimum may result in libraries that do not represent the data accurately [14]. Degraded RNA samples can also have a substantial impact on the accuracy of expression quantification during transcriptome analysis [15]. Therefore, RNA with minimal degradation and high levels of integrity are preferred when possible. In some cases, the sample type can make isolation of high-quality RNA a challenge. For example, biopsy specimens that are formaldehyde fixed, paraffin-embedded (FFPE) are subject to crosslinking of RNA with other macromolecules which then dramatically decreases RNA yield and quality [16]. When this is the case, the analysis of these datasets should be adjusted to take the RNA quality into account [15].

Because quantity and quality of RNA are so critical to the accuracy of a transcriptome, multiple complementary methods are used for their measurement. There are three main techniques for measuring RNA quality: (1) ultraviolet (UV) spectroscopy, (2) fluorometry, and (3) size separation.

UV spectroscopy can be used to quantify and differentiate between types of macromolecules because of their characteristic absorbance wavelengths [17]. While this can easily distinguish between some types of macromolecules, such as nucleic acids and proteins, it can be much more difficult to distinguish between DNA and RNA, which absorb similar wavelengths. UV spectroscopy is therefore useful to determine the amount of protein contamination in your sample but will not provide a measurement of DNA contamination.

Fluorometry can provide a more accurate measurement of RNA quantity in a nucleic acid sample by differentiating between its DNA and RNA composition. In this method, the samples are treated with fluorescent dyes that bind specifically to the molecule of interest (e.g., double-stranded DNA, single-stranded DNA, or RNA). Measurement of the fluorescence of the sample is determined by a fluorometer and this value is used to calculate the concentration of the specific nucleic acid type in a sample.

Size separation is the standard method used to assess RNA sample quality. In a sample of total eukaryotic cellular RNA, 80% of molecules are ribosomal RNA (rRNA) [18]. In eukaryotes, this includes the 28S rRNA molecules that are part of the large ribosomal subunit, and the 18S rRNA molecules that comprise the small ribosomal subunit (23S and 16S in bacteria). If the 28S and 18S rRNA molecules are intact, the overall integrity of all RNA molecules in the sample can be inferred to also be of high quality. To examine this, a small portion of the RNA sample is run through a gel matrix in a capillary tube to separate all of the RNAs by size, and the size distribution of the sample is measured. In a high-quality RNA sample, this will yield two large peaks that correspond to the 18S and 28S rRNAs (Fig. 5.3). If the ratio of the quantity of 28S to 18S rRNA is at least 2.0, the sample is considered to be of high quality. This ratio is used to calculate a value known as the RNA integrity number (RIN), which ranges between 1.0 (RNA is completely degraded) and 10.0 (RNA is completely intact) [19].

Once a sample has been determined to have sufficient quality to use for sequencing, the RNA sample is enriched for the RNAs that are of interest in the study. Because rRNA makes up the vast majority of the total RNA content in cells, preparation of an unenriched RNA sample will result in nearly all of the sequencing data representing these molecules. Several methods are available for enrichment of a sample of RNA, including selection for polyadenylated (polyA) RNA species, depletion of rRNA, or cDNA capture.

For mRNA-seq, a commonly used enrichment method is polyA selection. mRNAs (in both eukaryotic and bacterial cells) are polyadenylated: during transcription, a string of adenine nucleotides is added on to the $3'$ end of the RNA molecule [20] (Chap. 6,

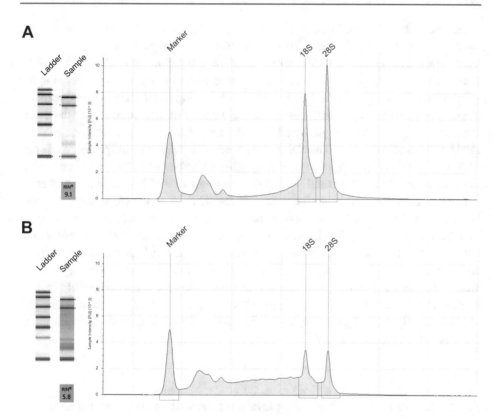

Fig. 5.3 Using size selection to assess RNA sample quality. Example traces showing the size distribution for RNA samples with (**a**) high integrity and (**b**) low integrity. Images of the size separation of the sample compared to ladder are shown on the left, and densitometry traces indicating the amount of fluorescence signal detected across the size range are shown on the right. The samples were analyzed using an Agilent 4150 TapeStation platform

mRNA). In this method, the 3′ polyA tails found on mature mRNAs are targeted for selection using magnetic beads coated with polyT oligonucleotide probes. These probes form complementary base pairs with the polyA tails of the mRNA molecules, allowing them to be pulled out of solution on a magnetic rack. RNAs that do not contain a polyA tail, such as rRNA, will be washed away and removed from the sample, quickly paring down the type of RNA to be sequenced. An alternative enrichment method specifically targeting rRNA molecules is ribosomal depletion, which is described in more detail in the section on sequencing long noncoding RNAs below.

Although polyA enrichment is an efficient way to enrich for mRNAs, it introduces notable biases into a sample, particularly the exclusion of any transcript that is not polyadenylated, which includes some types of long noncoding RNAs [21]. Additionally, because the mRNAs are selected for their 3′ end, any degradation or fragmentation of transcripts in the sample will cause the 3′ end of the transcript to be overrepresented relative to the rest of the molecule in the final sequencing library. This makes the integrity of the input RNA sample critical to the success of the final sequencing library.

Enrichment for an mRNA-seq experiment can also be performed using cDNA capture. This method is highly specific and can only be used when the sequences of interest in the sampled organism are known in advance. Rather than using polyT oligonucleotide probes, magnetic beads are coated with customized sequence that will complement the target sequences. This method is typically performed after the RNA sample has been used for cDNA synthesis (see below). The cDNA is allowed to base pair with the probes, while any molecules that do not match are washed away. This method enriches for specific transcripts to be sequenced, making it useful for focusing on a small set of known genes of interest. For example, cDNA capture has been used in the study of gene expression in different types of human cancer cells. Many mutations in coding sequences that contribute to cancer have been previously identified, and cDNA capture has been used to identify expression levels for the mutated alleles in cancerous cell types [22]. cDNA capture is a useful enrichment method for this type of experiment because it allows for targeting specific sequences, removing all background noise and homing in on a small number of genes of interest. In addition, cDNA capture can be useful when the genes of interest are expressed at low levels. By enriching the samples for these genes, they can be more readily identified and quantified in the sample [23].

5.2.2 Constructing Sequencing Libraries: Strandedness, Multiplexing, and Amplification

Once a high-quality RNA sample has been enriched for the RNA species of interest, the next step is to construct a sequencing library. Library construction is a multi-step process (Fig. 5.4) in which the RNA is fragmented and used as a template for cDNA synthesis. The cDNA molecules are then modified to contain adaptor sequences that are required for the sequencing reaction.

Although great care is taken during RNA isolation to ensure that samples are not degraded, the first step of mRNA-seq library preparation is to fragment the mRNA molecules to ~600 bp pieces (Fig. 5.4a) using physical, chemical, or enzymatic methods [24, 25]. This step is important for the quality of the sequencing library, as the final sequences generated during the sequencing reaction will be of a much shorter length than most mRNA molecules ($\leq$300 bp). In addition, the length of molecules in the library can introduce bias at several points during library construction, including during cDNA synthesis and PCR enrichment. By fragmenting the RNA into a uniform size distribution, these biases can be reduced [10]. Another advantage provided by fragmentation is the reduction of secondary structures that would inhibit efficient library construction. These structures, like hairpins or clovers, form through intramolecular base-pairing and are less likely to form in shorter RNA molecules.

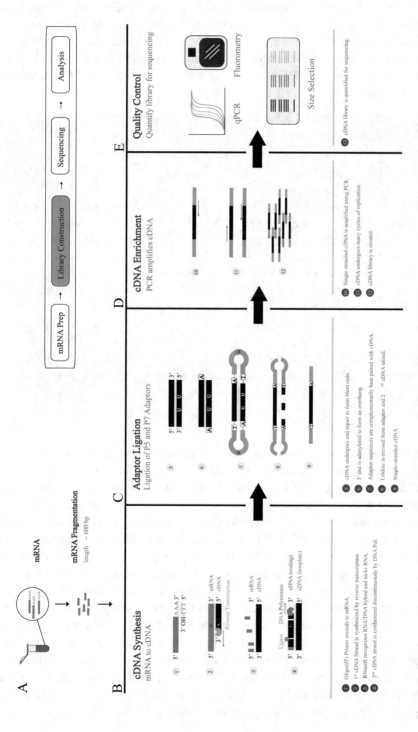

Fig. 5.4 Creating a stranded mRNA-seq library for SBS. Preparing an mRNA-seq library for SBS requires (**a**) mRNA fragmentation, (**b**) cDNA synthesis with dUTP-labeling of the coding strand, (**c**) sequencing adaptor ligation and removal of the second synthesized cDNA strand, (**d**) amplification of the cDNA library by PCR, and (**e**) quantification of the cDNA library by qPCR, fluorometry, and size selection

After fragmentation, RNA is used as a template to produce cDNA. cDNA synthesis provides an advantage because of the increased stability of a DNA molecule relative to RNA, allowing for easier handling and storage of the sample. In addition, SBS technology relies on DNA-based methods like Polymerase Chain Reaction (PCR—Box 5.1). Creation of a cDNA library from the RNA sample is therefore required for the downstream sequencing process.

Box 5.1 Polymerase chain reaction

Polymerase Chain Reaction (PCR) is a foundational method in molecular biology to amplify a targeted sequence of DNA (Fig. 5.5). DNA replication is carried out in vitro by combining a template DNA sequence, short oligonucleotide primer sequences that are synthetically produced to target a sequence of interest, free nucleotides, and a thermostable DNA polymerase. The reaction is subjected to 25–35 cycles of three steps. (1) The reaction is heated to a high temperature (95 °C) to *denature* the template DNA, or break the hydrogen bonds between base pairs. The DNA polymerase used in this reaction was isolated from the thermophilic bacteria *Thermus aquaticus* and can withstand this high temperature without itself being denatured. (2) The temperature is reduced (45–65 °C) to allow base pairs to form between the template DNA and the primer sequences. (3) The temperature is increased (72 °C) to provide optimal conditions for DNA synthesis. During this step, DNA polymerase uses the free 3′ hydroxyl of the primer to initiate synthesis of a new DNA strand that is complementary to the target DNA sequence. With each repetition of this cycle, the number of target sequences in the sample is doubled, resulting in 2^n copies of the target sequence, where n equals the number of cycles performed.

A critical consideration during cDNA library construction is whether your sample will be stranded or unstranded. When creating a stranded library, the information regarding the strand of DNA that was transcribed is maintained. During transcription, the DNA strand whose sequence is identical to the transcript is known as the coding strand, while the DNA strand whose sequence was used for complementary base-pairing by RNA polymerase is known as the template strand (Chap. 10, Transcription?). Stranded library preparations can therefore differentiate between transcription that took place to create a sense strand of RNA that matches the coding strand for a known gene, or transcription that took place to create an antisense strand of RNA that matches the template strand of a gene. Antisense transcription has been shown to play important roles in transcriptional regulation [26], making preservation of the strandedness of a library critical to the investigation of these functions. The increased amount of information provided by stranded libraries has led to them becoming standardized in the field for mRNA-seq library preparation protocols [10, 27].

To make a stranded cDNA library (Fig. 5.4b), a pool of random short oligonucleotides are added to the RNA samples, which can anneal to the RNA and act as primers by providing an available hydroxyl group on the 3′ carbon of the last nucleotide of the

Polymerase Chain Reaction (PCR)

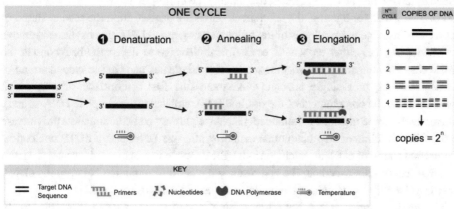

Fig. 5.5 Polymerase chain reaction (PCR). The events of one cycle of PCR are shown (denaturation, annealing, elongation). Amplification of a targeted DNA sequence by PCR is exponential, with the final number of copies equal to approximately 2^n, where $n =$ the number of PCR cycles

oligonucleotide. The enzyme reverse transcriptase uses this free 3'-OH for incorporation of nucleotides to synthesize a new cDNA strand from the RNA template in the 5' to 3' direction. This first cDNA strand corresponds to the template DNA strand that is complementary to the transcript sequence. After reverse transcription, the first strand of cDNA remains bound by complementary base-pairing to the RNA template, creating an RNA/cDNA hybrid. The enzyme RNaseH, which recognizes RNA/DNA hybrid molecules, is then used to create nicks in the RNA molecule. DNA polymerase I from the bacteria *Escherichia coli* uses the broken RNA molecule as primers to synthesize a second strand of cDNA, complementary to the first that reflects the sequence of the coding DNA strand. The second strand created through this method is discontinuous. *E. coli* DNA ligase is used next to create covalent phosphodiester bonds that complete sugar-phosphate backbone of the second strand of cDNA. During second-strand synthesis, DNA polymerase I is provided with a pool of nucleotides that includes deoxyuridine triphosphate (dUTP) rather than deoxythymidine triphosphate (dTTP). This marks the second strand of cDNA as distinct from the first, which will provide the strandedness of the library.

Next, the double-stranded cDNA molecules are modified to facilitate their use in a sequencing reaction (Fig. 5.4c). The cDNA first undergoes end repair using T4 DNA polymerase, *E. coli* DNA polymerase I, and T4 polynucleotide kinase to generate molecules with blunt and phosphorylated 5' ends before adding single 3'-A ends of the molecules. This overhang allows for complementary base-pairing with a 5'-T overhang found on adaptor sequences that are added to the ends of each cDNA fragment. The adaptors contain known DNA sequences that will be used during the sequencing process and are added to the cDNA molecules as hairpin loops including a uracil nucleotide in the center. The hairpin loop structure increases ligation efficiency and decreases adaptor dimerization events, as there is less steric hinderance during ligation.

Following adaptor ligation, the uracil nucleotides are excised from the second strand of cDNA and from the hairpin loops of the adaptors using uracil DNA glycosylase to remove the uracil base and either enzymatic or chemical cleavage to digest the backbone at the abasic site. This linearizes the adaptor sequences and creates gaps in the second strand of cDNA, resulting in a single intact cDNA strand (the first synthesized strand), which corresponds to the complement of the original RNA molecule (the template DNA strand). Alternatively, some methods for stranded mRNA-seq library construction use a polymerase in the PCR enrichment step below that cannot synthesize DNA past a dUTP nucleotide. The second strand of cDNA is excluded as a PCR template in this method because it cannot be amplified. In either method, the end result is a clear differentiation between the cDNA strands to identify the template and coding strands of DNA for each mRNA.

The single-stranded cDNA library remaining is then enriched using PCR (Fig. 5.4d). Extension primers that complement the adaptor sequences are used to amplify the cDNA. These primers contain additional sequences required in the SBS sequencing reaction, known as P5 and P7 sequences. The PCR enrichment results in cDNA library fragments with a P7 sequence on the 5' end of the cDNA strand that corresponds to the coding strand of DNA, and a P5 sequence on the 5' end of the cDNA strand that corresponds to the template strand of DNA. This provides the directionality required to determine the strand-edness of each sequence generated in the sequencing reaction. In addition, the P7 sequence also contains a barcode that is unique to a sample, allowing for multiplexing of samples during the sequencing reaction. In a multiplexed sequencing reaction, multiple samples are pooled and sequenced simultaneously with the sample barcodes allowing the sequences from each sample to be separated following the sequencing reaction. Once PCR enrichment is complete, the library is carefully quantified using quantitative PCR (qPCR), fluorometry, and size selection (Fig. 5.4e) and is loaded onto the sequencing platform.

5.2.3 Generating the Transcriptome: Bridge Amplification and Sequencing-by-Synthesis

Sequencing of a cDNA library for mRNA-seq using SBS technology [28] occurs in two parts (Fig. 5.6). First, each cDNA molecule in the library is amplified on a solid-state flow cell, resulting in clusters of identical sequences that will serve to increase the sequencing signal. Second, cycles of reversibly terminated nucleotide incorporation occur. This allows for fluorescently labeled nucleotides to be added one at a time, with each incorporated nucleotide detected as the strand is synthesized. After 50–300 cycles of nucleotide incorporation, this process is often repeated for sequencing from the other end of the cDNA fragment to produce paired-end data.

SBS occurs on a flow cell, which is a glass slide containing multiple channels, also referred to as lanes. During the sequencing reaction, reagents needed for the sequencing

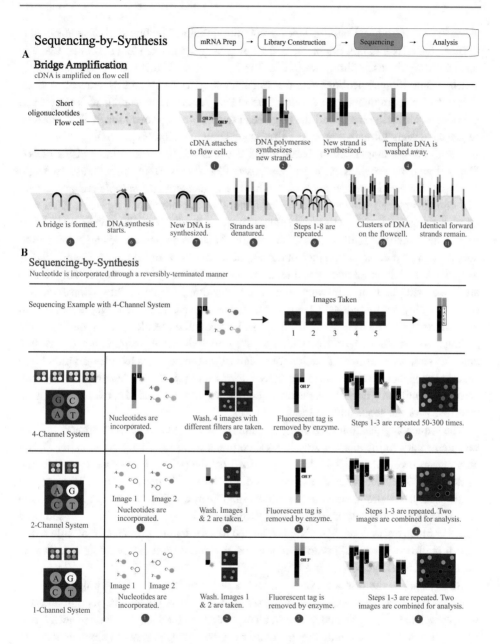

Fig. 5.6 Sequencing-by-synthesis (SBS). SBS begins with (**a**) bridge amplification to create clusters of identical library fragments on a flow cell. After cluster generation, (**b**) sequencing-by-synthesis begins with cycles of reversibly terminated nucleotide incorporation and base-calling that occurs using 4-channel, 2-channel, or 1-channel chemistry

chemistry are introduced into the lanes using microfluidics. The first step of sequencing is to attach the cDNA library fragments to the flow cell and generate identical copies that will form a cluster for each library fragment. The flow cell surface is coated with covalently attached short oligonucleotides that are complementary to the P5 and P7 sequences found on each cDNA fragment (see above). The cDNA library is denatured and allowed to hybridize with these oligos, and the oligos can then serve as primers for PCR to generate sequence clusters through a process called bridge amplification (Fig. 5.6a).

During bridge amplification, base-pairing between the flow cell oligo and the cDNA library fragment provides a free $3'$ hydroxyl that allows DNA polymerase to begin synthesis. DNA polymerase continues synthesis to create a full complementary strand. The double-stranded molecule is denatured, and the template cDNA molecule is washed away, leaving the newly synthesized strand that was extended from the covalently attached oligo. The new strand folds over to form a "bridge" and the non-anchored end hybridizes to another complementary oligonucleotide bound to the flow cell surface. A new cDNA template strand is amplified, forming a double-stranded cDNA molecule. The strands are denatured, and the process repeats several times. Finally, the reverse strands (corresponding to the template DNA strands) are cleaved from the oligo sequence and washed away, leaving a cluster of identical forward strand molecules (corresponding to the coding DNA strands). The cluster undergoes the same sequencing reaction in the next step, and the simultaneous incorporation of identical fluorescent nucleotides amplifies the sequencing signal. Each fragment of cDNA forms its own cluster, leaving the flow cell coated with millions (or billions, depending on the sequencing platform) of clusters that are now ready for the sequencing reaction.

The sequencing reaction (Fig. 5.6b) is initiated through the addition of primers that are complementary to the adaptor sequences incorporated during library construction. DNA polymerase uses the free $3'$-hydroxyl provided by the primer to incorporate additional nucleotides. The free nucleotides provided to DNA polymerase for addition to the synthesized DNA strand contain two critical modifications. First, each type of nucleotide is tagged with a fluorescent dye (dTTP = green, dCTP = yellow, dATP = red, and dGTP = blue). Second, the $3'$ carbon is attached to a reversible terminator molecule rather than a hydroxyl group. This modification ensures that only a single nucleotide can be incorporated by DNA polymerase [29].

SBS occurs as the reaction cycles through a series of steps. (1) DNA polymerase and modified free nucleotides are added to the reaction, and DNA polymerase incorporates the next complementary nucleotide. The unincorporated nucleotides are washed away, and (2) the fluorescent color emitted by the cluster is recorded. (3) Enzymes are then added to the reaction to remove the fluorescent tag and the terminator molecule from the $3'$ carbon, providing a free hydroxyl that will allow for extension by DNA polymerase in the next cycle. These steps are repeated 50–300 times, and the order of fluorescent color emissions for each cluster on the flow cell is interpreted as a DNA sequence.

Notably, some sequencing platforms that use SBS (such as Illumina NextSeq and Illumina iSeq) have adjusted technology to use a two color (2-channel) or one color

(1-channel) system, respectively, rather than 4-channel to increase the efficiency of the sequencing run (Fig. 5.6b). In a 4-Channel system, a different color is used to identify each type of nucleotide, each requiring a separate filter for detection of its color, as described above. In 2-Channel sequencing, dTTP fluoresces green, dCTP fluoresces red, dATP fluoresces both green and red, and dGTP lacks fluorescence [30]. This combination requires only two filters, green and red, to detect all four nucleotides. In contrast, 1-channel sequencing methods label dTTP with green fluorescence, dATP with green fluorescence that can be enzymatically removed, dCTP with a linker group, and dGTP without fluorescence. Following nucleotide incorporation, an image is taken. The dATP fluorescence is then removed while fluorescent labels are added to the linker group attached to dCTP and a second image is taken. The pattern of loss or maintenance of the fluorescence state is used to infer the nucleotide that was incorporated during that cycle [31]. Using 1-channel SBS still requires two images during each cycle of sequencing, but needs only a single filter. Decreasing the filter number required for sequencing reduces the processing of the colors during each cycle of the sequencing run. Although the 1- and 2-channel systems are more prone to errors due to the increased probability of misinterpreting a fluorescent signal, the overall error rate for sequencing using this method remains very low.

Following the 50–300 cycles of nucleotide incorporation and fluorescence detection, a collection of single-end sequences have been collected, with one sequencing read (the raw sequence produced during the sequencing reaction) generated per cDNA library fragment. However, the sequencing process can be repeated to produce a second read for each fragment, beginning from the opposite end of the cDNA fragment, and sequencing in the opposite direction. To accomplish this, once the first read has been generated, bridge amplification is repeated and the forward strand is cleaved and washed away from the flow cell. New sequencing primers are added and the reverse strand is used as a template for SBS. Samples that undergo this additional round of sequencing now have paired-end reads, with each cDNA molecule having two sequencing reads that reflect each end of the fragment [32].

Paired-end sequencing provides several advantages over single-end data. Paired-end reads provide not only twice the amount of sequencing data, but also allow for spatial inferences. The cDNA fragments that make up the sequencing library are of known approximate length due to the RNA fragmentation step during sequencing library preparation. Therefore, the length of the unknown sequence that separates the two reads can be used to aid in assembly of transcripts during analysis (see below). Whether or not this additional information is required for mRNA-seq depends on the nature of any particular experiment, the availability of genome or transcriptome sequences for the organism of interest, and the planned data analysis.

5.2.4 Making Sense of the Data: Transcriptome Assembly and Differential Expression

Analyzing the very large amount of sequencing data generated in an mRNA-seq experiment progresses through a well-established workflow requiring specialized software (Fig. 5.7). The steps of mRNA-seq analysis typically include quality control and read processing, transcriptome assembly, and differential expression analysis [11, 33]. Once these steps are completed, a full picture of the genome-wide transcriptional changes in protein-coding sequences will be created.

The first steps of mRNA-seq analysis are focused on assessing the quality of the sequencing data and performing any necessary editing of the sequencing reads (Fig. 5.7a) [14]. The raw sequencing data is a large text file that contains both the nucleotide sequence of each individual sequencing read, and a quality score for every base call made by the platform during the sequencing reaction. The quality score reflects the statistical confidence in the base call for each individual nucleotide in every sequencing read that was generated. To assess the quality of the overall sequencing run, the attributes of the sequencing reads are analyzed, including examining the quality scores of the reads, the nucleotide composition (such as the GC content), the lengths of the sequence, and the frequency of identical reads in the sample. Filtering of the reads may be required at this point if any major issues are observed with the sample to remove low-quality sequences before additional analysis is performed. Reads are further processed by identifying and trimming any adaptor sequences before performing further analysis.

The next portion of mRNA-seq analysis is to use the short sequences produced by the sequencing reaction to assemble full-length transcripts (Fig. 5.7b). Depending on the experiment and the resources available for the organism being studied, the approach for this step can fit into one of two methods: reference-based or de novo assembly [34].

Reference-based assembly requires a previously generated genome or transcriptome assembly. For this method, the short sequencing reads are "mapped" to a reference sequence (genome or transcriptome). Mapping entails aligning the reads to the reference where the sequence is an identical match. After mapping, transcripts are assembled from overlapping reads. Challenges can arise for reference-based transcriptome assembly in organisms that contain repetitive genomic regions. In these cases, reads can map to multiple locations in the reference sequence, which makes the true transcriptional location ambiguous. An additional challenge occurs in organisms that contain introns and use alternative splicing. In this case, multiple transcript isoforms may exist for a single gene. Mapping software identifies splicing events using reads that span the junction between exons and therefore map to regions that flank intron sequences.

De novo assembly is required when a reference sequence is not available or is not of sufficient quality for the organism under study, which is most likely to be the case in studies of non-model organisms. De novo ("from nothing") assemblies build transcripts with no prior knowledge of the transcriptome's contents. Since no reference sequence is used, de novo assembly programs use de Bruijn graphs, which are a way of mapping paths through a

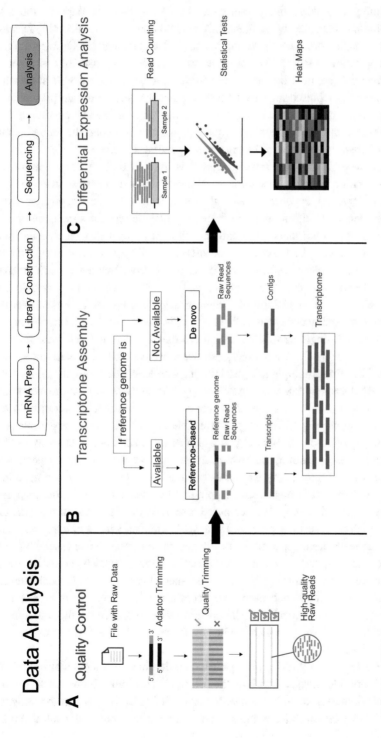

Fig. 5.7 mRNA-seq data analysis workflow. Analysis of mRNA-seq data requires (**a**) quality control steps to remove sequencing adaptors and low-quality sequencing reads and (**b**) assembly of the short reads into full-length transcripts using a reference (reference-based) or from scratch (de novo). In many experiments, (**c**) differential expression analysis is performed to quantify and compare levels of gene expression between two or more samples

sequence assembly to predict likely transcripts based on short overlaps in the read sequences. Paired-end data greatly facilitates this de novo transcriptome assembly because of the spatial information provided by having two reads from the same cDNA fragment that is of known approximate length. The final result of a de novo assembly is a set of contiguous sequences (contigs) that were built from the sequencing reads. If successful, there will be one contig representing each transcript. Since the only sequences present in an mRNA-seq dataset are polyA transcripts expressed under the experimental conditions, this assembly represents the set of coding sequences expressed under the experimental conditions rather than the entirety of the organism's genome sequence.

Whether reference-based or de novo assembly is used to build a transcriptome, an important consideration is the transcriptome variability that will occur between samples. A well-designed mRNA-seq experiment will always include biological replicates for a particular experimental condition, and will frequently include replicates generated under different experimental conditions. Although the biological replicates receive identical treatment between samples, biological and technical variability will result in differences in the assembled transcriptome between replicates. In addition, changes in experimental conditions may result in specific transcripts or isoforms only being represented in one condition or another due to specificity of their expression conditions. To ensure that the final transcriptome assembly is inclusive of all transcripts, the reads from all samples can be pooled to generate one single transcriptome that can then be used as a reference for read mapping or annotation [11]. Alternatively, software is available that will merge the transcript assemblies from multiple samples into one single holistic assembly [35].

For many mRNA-seq experiments, the next step of the analysis will require examining differential expression, or the changes in expression that occurred as a result of the experimental conditions (Fig. 5.7c). A powerful feature of mRNA-seq experiments is that the sequences produced not only tell you which genes are expressed, but also provide quantitative information regarding the level of expression occurring for every gene. The amount of expression is inferred by the number of reads that originate from a gene in a sample. This is determined through read mapping and counting the reads that map to a particular gene or transcript. For reference-based assembly, this information may already be captured through the assembly process. De novo assemblies will require an additional step, in which the reads are mapped to the newly generated transcriptome assembly.

Once the reads are mapped and counted, the quantity of reads for each gene or transcript is compared between the samples from each experimental condition. Statistical tests are performed to assess the likelihood each observed change in expression is greater than would be expected by random chance. These tests take into account the variation in expression observed between biological replicates, as well as the level of expression of a particular gene or transcript.

Using mRNA-seq to study differential gene expression has been a popular use of this technology as it provides insight into how the cell modulates transcriptional activity in a variety of cellular contexts. In addition, a wide range of experiments can also be designed using differential expression to learn about gene regulation and transcriptional networks.

By making controlled changes to the genome, such as gene deletions or insertions, and comparing mRNA expression in these cells to wild-type cells, the effects of these changes can be examined genome-wide. Experimental manipulation of the genome is particularly useful when it comes to understanding the effects of transcription factors on the expression of other genes.

The complexity of the transcriptome beyond differential expression can also be investigated using mRNA-seq methods [33]. The high resolution of transcript sequence information provided through mRNA-seq can allow for determination of allele-specific expression at heterozygous loci in the genome. This information can also be integrated with variation in gene expression levels across samples to identify expression quantitative trait loci (eQTL), which are sites that provide genetic contributions to this variation.

5.3 Beyond mRNA-Seq: Other High-Throughput Sequencing Applications for RNA

Focusing an experiment on RNA molecules that encode for protein products can be a powerful way to gain insights into cellular responses and pathways. However, other applications have also been developed that focus on other pieces of the transcriptome, such as long and small noncoding RNAs, or nascent transcripts. In addition, library construction has been adapted to study the interactions between RNA molecules and other cellular components, such as the ribosome and RNA-binding proteins. These tools expand the repertoire of sequencing-based methods that provide a picture of RNA function.

5.3.1 Long Noncoding RNA

Long noncoding (lncRNAs) are RNA transcripts greater than 200 base pairs long that perform a function as RNA molecules rather than encode for proteins. LncRNAs perform a wide range of important cellular functions, including tethering transcriptional machinery to DNA, increasing mRNA stability, and regulating transcription of other genes [21]. Preparing a sequencing library that includes lncRNAs is very similar to the library preparation procedure for mRNA-seq described, but can include some important modifications.

One critical difference between preparing an mRNA-seq library, and a total RNA-seq library that includes lncRNAs is in the enrichment method. In mRNA-seq, a sample of RNA was enriched for mRNAs that have a polyA tail. Although some lncRNAs are polyadenylated, many are not and would therefore be lost during library preparation using this method. Therefore, an alternative enrichment method known as rRNA depletion is often used instead. rRNA depletion is a more inclusive enrichment method that selectively *removes* one class of RNA (rRNA) from the sample rather than selectively *retain* one particular class of RNA molecule. During the enrichment step, the samples are mixed with magnetic beads coated with oligonucleotides that complement rRNA sequences found in the organism of interest. Once the rRNAs have hybridized with the oligos on the beads, the beads are pulled down on a magnet and discarded while the rest of the sample is kept for library preparation. This leaves behind a diverse pool of transcripts for sequencing, regardless of polyadenylation status.

Although an effective and commonly used enrichment method, there are some important considerations when using rRNA depletion for sample enrichment. Kits that have been manufactured for this method are typically made for rRNA sequences found in model organisms. Although these sequences are well-conserved, they may not effectively remove all rRNA molecules in non-model organisms when nucleotide differences in these sequences exist between species. In addition, cross-reactivity between sequences may occur if a sequence found in the rRNA probe matches a sequence elsewhere in the transcriptome. This would result in inadvertent removal of the transcript from the sample and its absence from the final sequencing data. To reduce these concerns, rRNA depletion can be performed through selective degradation rather than with magnetic beads. For selective degradation, a pool of longer complementary DNA oligos are added to the sample that will anneal to the entire length of the targeted rRNA. The RNA:DNA hybrids are then targeted for enzymatic degradation by RNaseH and DNaseI [36]. This method may be more effective for enrichment of a sample from a non-model organism whose RNA may not be removed entirely by a commercially available depletion kit. Additionally, this method can be adapted to target not only rRNAs, but also other highly expressed transcripts that perform constitutive functions in the cell (i.e., transcripts for housekeeping genes). This selective removal allows transcripts expressed at very low levels, such as many lncRNAs to be more easily discovered in the sample [21].

Following the enrichment step, the rest of the sequencing library construction occurs in the same way as for mRNA-seq. In the case of sequencing lncRNA, creating a stranded library is absolutely essential due to the antisense transcription used to produce many of these molecules [37]. Transcriptome assembly is also very similar to mRNA-seq. However, after transcriptome assembly is complete, candidate lncRNAs require further computational analysis of their coding potential and/or experimental evaluation of their structure and function [38].

5.3.2 Small Noncoding RNAs (miRNAs, tRNAs, etc.)

Small noncoding RNA (sncRNA) includes transcripts that do not encode for proteins and have a shorter length (often less than 100 nucleotides). This includes RNAs that function in regulating the expression other genes, such as microRNA (miRNA) and small interfering RNA (siRNA), and RNAs that are involved in maintaining genome integrity, such as piwi-interacting RNA (piRNA) (Chap. 8, ncRNA) [39]. RNA-seq methods have facilitated sncRNA discovery and quantified their expression under a variety of conditions, providing insights into their roles in critical cellular processes and in the development and progression of human disease [40].

Performing a sequencing experiment targeting sncRNAs differs from mRNA-seq in the enrichment stage of the library preparation protocol. In the case of sncRNAs, the enrichment methods focus on selecting for RNA species based on their size. Because coding mRNA molecules tend to be longer than 500 nucleotides, selection is performed for RNA molecules that are shorter than this length. This can be done by separating the RNA sample by size on a polyacrylamide gel, excising the desired size range of sample, and extracting the RNA. Methods and specialized equipment have been created to automate this process, which makes it much more precise and increases the RNA yield from the extraction. Following the enrichment, no fragmentation step is required, as the transcripts are already at a length that is short enough for the sequencing reaction.

In some cases, an alternative method of enrichment is used for sncRNA library preparation. Because many sncRNAs require interactions with specific RNA-binding proteins to carry out their functions, these RNA-protein interactions can be used to enrich for specific types of sncRNA. For example, miRNA bind to proteins in the Argonaute family to target specific mRNA molecules for degradation or inhibit their translation in numerous eukaryotic organisms [41]. To enrich for miRNAs in a sample, antibodies that recognize the argonaute proteins can be used for immunoprecipitation that will pull down both the argonaute proteins and any RNA molecules that they interact with. Although this method is much more specialized than size selection, and would not be useful for examining the entire small noncoding transciptome, it allows for focus on a particular subset of sncRNAs and may allow for discovery of molecules expressed at low levels that would be obscured by other enrichment techniques.

Generating a cDNA sequencing library for sncRNAs also differs from mRNA-seq library preparation methods. For sncRNA library preparation, adaptors are ligated to the 3' end of the RNA molecules first [42]. Primers are then added that hybridize to the 3' adaptor before ligating adaptors to the 5' end of the RNA molecules. The sample is then enriched for RNA molecules that contain the primer and both adaptors before cDNA synthesis and PCR enrichment is performed. These modifications increase the yield of molecules represented in the final sequencing library. However, a challenge of sncRNA library preparation methods is the bias that can be introduced. Molecules that are expressed at low levels may not have accurate quantitative representation in the final library. In addition, sncRNAs in some organisms contain posttranscriptional modifications that make adaptor ligation less efficient. Adjustments to the adaptor sequences and reaction conditions can help mitigate these biases for some types of sncRNAs [42].

Analysis of a sncRNA sequencing library is very similar to the mRNA-seq workflow. Following quality control, read mapping, and differential expression analysis, additional work may be required to further investigate the functional implications of the data. For example, miRNA sequences of interest may be analyzed to predict their mRNA target sequences. Conclusions from these types of analyses require further investigation and validation through experimental work.

5.3.3   Investigating RNA Biology: Other Applications of RNA Sequencing

In addition to studying gene expression dynamics, sequencing methods can be used for large scale examinations of cellular mechanisms that involve RNA. These types of methods focus on a variety of aspects of RNA biology, including RNA structure, transcription, translation, and other mechanisms that involve RNA–protein interactions. A brief sampling of these methods is provided below, although many more continue to be developed to pursue a variety of questions related to RNA.

Studying Translation Using Ribo-Seq Although mRNA-seq methods will provide information about the transcripts present under the conditions of a particular experiment, it does not allow direct inferences about the proteins being synthesized. Regulation of the timing and rate of translation are critical processes in the control of gene expression by the cell. To examine translation dynamics, all mRNA molecules that are interacting with ribosomes are sequenced (Ribo-Seq) providing a snapshot of all mRNAs being actively translated [43]. To accomplish this, cells are treated with the antibiotic cycloheximide to cause translation to stall. The treated cells are then lysed, RNA is extracted, and RNases are used to enzymatically degrade all RNA that is not bound to the ribosome. The mRNA-ribosome complexes are purified using centrifugation or chromatography, and mRNA is purified from the ribosomes for use in creating cDNA sequencing libraries. The sequences generated provide a ribosome profile that shows not only *which* mRNA molecules were being translated, but *which part* of a particular mRNA was being translated. It also provides quantitative information that can be used to infer rates of translation for each mRNA. Ribo-seq data can be used to identify novel open reading frames (ORFs), alternative translation initiation sites in known ORFs, and can be used in combination with mRNA-seq to examine transcript stability and post-transcriptional gene regulation.

Studying Transcription Through Nascent RNAs The mRNA-seq methods described above reflect the steady state of RNA found in the cell, or the summation of RNA that is produced and degraded in a particular condition. However, RNAs that are inherently unstable, such as RNAs that are produced at enhancer sequences (enhancer RNAs, or eRNAs) that have functions in regulating enhancer activity, are not easily examined this way. In addition, the dynamics of the process of transcription are not reflected in mRNA-seq datasets. Insights into RNA polymerase pausing, transcriptional termination, and RNA modifications that occur concurrently with transcription (capping, splicing, and polyadenylation) (Chap. 6, mRNA; Chap. 10, transcription) can be investigated by examining nascent RNAs, which are those molecules that are in the process of being transcribed [44].

Several approaches can be taken to examine nascent transcription using high-throughput sequencing methods [6, 44]. (1) RNA that is associated with chromatin can be isolated using salt washes (caRNA-seq). These samples would include not only nascent RNAs, but also spliceosomal RNAs and any lncRNAs that functionally associate with DNA. (2) RNA polymerase II, which is responsible for transcription of mRNAs, small nuclear RNAs, and miRNAs, can be tagged with a small epitope. Following isolation of chromatin, the epitope tag can be used to immunoprecipitate RNA polymerase II and any RNAs with which it is associated. The RNAs are then purified from RNA polymerase II and used to create a sequencing library (mNET-seq). (3) Run on methods can be used, in which transcription in the cell is halted through drug treatment or freezing temperatures, nuclei are isolated, and transcription is allowed to resume in vitro with nucleotide analogs such as 5-bromouridine 5′-triphosphate (BrUTP). BrUTP-containing RNAs, which represent those that are newly transcribed, can then be enriched in the final sample by targeting the analogs for immunopurification (GRO-seq). (4) In a manner similar to a run on method, newly produced RNAs can be examined in vivo by providing cells with labeled nucleotides in their media and allowing them to be incorporated for a defined period of time. Following RNA isolation, the samples can then be enriched for these nascent transcripts by targeting the labeled molecules through immunopurification (TT-seq, TimeLapse-seq, or SLAM-seq).

Studying RNA–RNA and RNA–Protein Interactions RNA function can depend on its interactions with other RNA molecules (e.g., miRNA binding to the 3′ UTR of target mRNAs) or with RNA-binding proteins. A variety of methods have been developed to examine these interactions using high-throughput sequencing technologies to define a cellular "interactome" [6]. RNA–RNA interactions can be identified by using biotinylated psoarlen, which intercalates into RNA–RNA hybrid molecules and crosslinks the interactions (Sequencing of Psoralen crosslinked, Ligated, And Selected Hybrids, or SPLASH). The crosslinked molecules can be purified using streptavidin to pull down the biotinylated molecules. Following purification, the RNA hybrids are fragmented and ligated to join the interacting molecules into a single strand that is used for library preparation and sequencing. Intermolecular RNA-RNA interactions are inferred during analysis based on the sequencing reads that contain more than one RNA species.

To examine interactions between RNA and proteins, immunoprecipitation with antibodies targeting a protein of interest is performed either without crosslinking or following ultraviolet (UV) crosslinking of the cells (RIP-seq and CLIP-seq, respectively). UV crosslinking will form covalent bonds between RNA and protein, but will not crosslink protein–protein interactions, which reduces the noise in the sample. Following immuno-precipitation, the RNA is purified from the protein and used as the input for library preparation and sequencing. Analysis of these datasets can reveal all RNAs that interact with a particular protein and can be used to identify the sequence motifs found in the RNA molecule that are recognized by the protein.

Studying RNA Structure RNA molecules can form secondary structures through intra-molecular base-pairing interactions, as well as tertiary structures. These structures are critical to their functions and interactions with other molecules in the cell (Chap. 2, Architecture of RNA) [45]. The "structurome" can be established through high-throughput sequencing methods [46]. To determine the unstructured, single-stranded regions and structured, double-stranded regions of RNA molecules, selective enzymatic digestion of single-stranded or double-stranded RNA is performed (PARS or FRAG-seq). The remaining RNA is used for library preparation and sequencing. Performing the experiments to create ssRNA and dsRNA sequencing libraries in parallel provides a full picture of the RNA structurome under the experimental conditions. Chemical mapping can be used as an alternative to nuclease treatment, in which chemical probes can be used to mark structured or unstructured RNA and targeted to enrich the final library prep (SHAPE-seq).

5.4   The Present and Future of Transcriptomics

Short-read sequencing-by-synthesis methods have served as a powerful means to generate quantitative surveys of gene expression for many types of RNA across a diversity of organisms. Now that well-established protocols are in place to conduct differential expression analysis and de novo transcriptome assembly, there is wide implementation of them in a range of experimental frameworks.

More recent developments have focused on expanding these transcriptome studies to incorporate new technologies that address some of the key limitations of short-read HTS sequencing applications to transcriptome work, as well as expand the technological toolbox for studying gene expression to include methods aimed at better understanding heterogeneity in gene expression and RNA modifications.

5.4.1   Improving Assembly: The Advent of Long-Read Sequencing Methods

Following the rise and success of short-read SBS methods developed by Illumina, a so-called "third-generation" or "next-next generation" of sequencing technologies has now emerged and expanded. Key innovations in these technology developments have included *increasing the length* of sequencing reads and *removal of pre-sequencing sample processing steps* to allow for more efficient and less biased data. In particular, the increased length of the sequencing reads allows for greatly improved de novo assembly of the transcriptome and enhances the detection of transcript isoforms with less ambiguity [47].

These new methods of sequencing differ from short-read SBS methods in two key ways: sequencing a *single-molecule* of DNA or RNA at a time and carrying out sequencing of a molecule *directly*. In particular, two methods have gained popularity, each having distinct sequencing mechanisms underlying their technology.

The first method of third-generation long-read sequencing is single-molecule real-time (SMRT) sequencing, developed by Pacific BioSciences (PacBio) [48]. To conduct transcriptome studies using this method, RNA is first converted into cDNA, and hairpin adpators are ligated to the ends of the cDNA molecules. This creates a circular single-stranded molecule that is used for sequencing. Following adaptor ligation, the cDNA library is loaded onto a SMRT cell, which is composed of small wells called zero-mode waveguides (ZMWs). Each ZMW contains a DNA polymerase enzyme fixed to the bottom. The DNA polymerase binds to the cDNA molecule and initiates DNA synthesis using the cDNA molecule as a template (Fig. 5.8a). During synthesis, the DNA polymerase is provided four differentially labeled fluorescent nucleotides that will emit a pulse of light once the nucleotide is incorporated into the synthesized strand. The series of pulses generated during strand synthesis are detectable by an imager and are interpreted as a DNA sequence.

There are two key features of SMRT sequencing that distinguish this method from the short-read SBS method implemented by Illumina [48]. First, sequences reflecting entire RNA molecules can be represented in the final dataset, as no RNA fragmentation is required during library preparation prior to cDNA synthesis. Second, no bridge

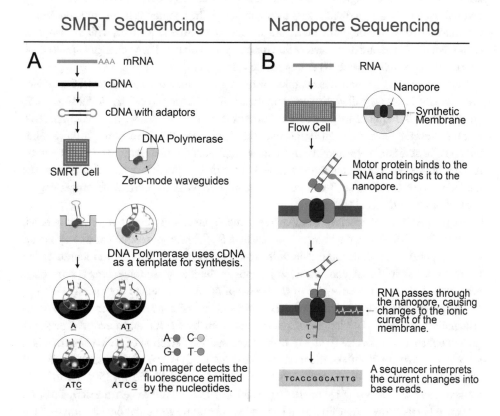

Fig. 5.8 Long-read sequencing methods. Long-read sequencing methods include (**a**) SMRT sequencing and (**b**) nanopore sequencing

amplification step occurs prior to SMRT sequencing; the fluorescent signal emitted by the single synthesized molecule in the ZMW is detected by the imager without the need for a cluster of identical molecules. Removal of the need for fragmentation and sequence amplification for SMRT sequencing reduces the errors and biases that can be introduced by these steps. In addition, the length of reads that are produced through SMRT sequencing is substantially increased at up to 25 kilobases per read. This increased length comes at a cost of total reads produced, however, as a typical run of SMRT sequencing yields 4,000,000 reads (compared to the up to 20 billion reads produced by the Illumina NovaSeq platform). In addition, the error rates in SMRT sequencing are higher than in SBS (~15% vs. ~0.1%, respectively).

The second method of third-generation sequencing marketed by Oxford Nanopore Technologies, high molecular-weight RNA molecules are sequenced directly without the need for cDNA conversion [49]. The sequencer contains a synthetic membrane with hundreds or thousands of nanopores, depending on the sequencing platform used (Fig. 5.8b). RNA molecules bound by motor proteins are brought to the nanopores. As the RNA is passed through the nanopore, it disrupts an ionic current that is formed by a preestablished voltage gradient across the membrane. The alterations to the ionic current as the molecule moves through the nanopore are detected by the sequencer, with the shape of each nucleotide creating a characteristic change to the current. These changes are used by the sequencer to infer base calls and generate a sequencing read.

The feature that distinguishes nanopore sequencing from both SBS and SMRT sequencing is that the bases are called *directly* from the RNA molecule. No cDNA synthesis or PCR amplification is required as part of the library preparation or sequencing reaction. This removes several forms of sequencing bias that have been observed through these other methods. In addition, detection of ionic current changes does not require an imager, which reduces the size of the sequencing equipment to as small as the palm of a hand. This makes the technology portable, so it can be brought into the field for immediate processing of clinical or environmental samples.

As with SMRT sequencing, nanopore sequencing results in much longer read lengths, lower total read output, and a higher error rate than SBS. Nanopore sequencing can yield up to 242 gigabases of sequencing data with read lengths limited only by the length of the RNA fragments in the sequencing library (reads in the megabase range length have been achieved). However, the error rate of the base-calling by this method is typically ~10%.

Long reads generated by third-generation sequencing methods are the basis for isoform-sequencing (Iso-seq). The length of reads sequenced by SMRT and nanopore methods greatly enhances the assembly of a transcriptome and the detection of splice isoforms in a sample. Because the length of the reads is on the order of several kilobases, which is well within the average length of an mRNA transcript in humans (3522 bp) [50], *Drosophila melanogaster* (3058 bp) [51], or yeast (~1250 bp) [52], a full-length transcript can be sequenced in its entirety as a single read and does not require computational assembly. This overcomes the challenges in transcriptome assembly from SBS data of identifying splice-junctions and in assembling alternatively spliced transcripts. Thus, with generation of a

SMRT or nanopore transcriptome, the library of transcript isoforms can be identified fully and without bias or ambiguity introduced through the computational methods of assembly [49].

In important limitation in long-read sequencing technologies is the increased error rate in the reads relative to short-read sequencing [47]. Although some forms of bias in sample generation are removed through these methods, others are introduced, and the reported error rate for PacBio SMRT (~15%) and Oxford Nanopore (~10%) sequencing methods is substantially higher than SBS (~0.1%). This can be addressed in the downstream analysis by increasing the *coverage* of the transcriptome (i.e., sequencing multiple reads per transcript), thus allowing for the representation of more correct than incorrect sequences and allowing a consensus sequence to be inferred. With SMRT sequencing specifically, this can also be mitigated through the generation of circular consensus sequences (CCS). CCS generation takes advantage of the hairpin adaptors/circular structure of the sequenced molecule. By continuing to synthesize DNA from the circular molecule, you can sequence the same molecule multiple times, generating a long read that can be chopped up and assembled into a consensus sequence that represents the fragment [48]. By generating redundancy in the sequencing reaction, you can identify individual nucleotide errors and remove them from your downstream analysis.

5.4.2   Examining the Epitranscriptome: Direct Detection of RNA Modifications

Posttranscriptional modifications to both noncoding and coding RNA molecules can impact their structure and function by influencing their stability, localization, and interactions with other molecules. Well-known modifications to RNA, such as N6-methyladenosine, 5-methylcytosine, 7-methylguanosine, pseudouridine, and adenine to inosine editing, can be detected and quantified through SBS methods [53]. However, using SBS results in indirect detection and requires extensive library preparation that often involves immunoprecipitation.

The development of long-read sequencing methods has led to the ability to sequence RNA molecules directly (Direct RNA-seq), which can allow for direct detection of posttranscriptional modifications. Nanopore sequencing has been used successfully to detect a variety of RNA modifications [54]. Just as the shape of the individual bases changes the ionic gradient across the membrane in specific ways, unique signatures are detected when a base is modified. Thus, as the molecule passes through the membrane, the bases can be read along with their modifications to generate not only the sequence of the RNA, but to also determine which bases have been modified, and with which particular modification.

SMRT sequencing has also been tailored to allow for the detection of RNA modifications [54]. By modifying the sequencing reaction to use reverse transcriptase instead of DNA polymerase for nucleotide incorporation, the RNA molecules can be sequenced directly without the need for cDNA library construction. The kinetics of nucleotide incorporation during SMRT sequencing are changed in consistent and predictable ways when the template contains modifications. The altered kinetics can be interpreted by the sequencer as particular posttranscriptional modifications. Although detection of nucleotide modifications using SMRT sequencing has more frequently been applied to genomic DNA sequencing, it has also been successfully applied to transcriptome sequencing.

5.4.3 Deciphering Heterogeneity: Transcriptomes from Individual Cells

The standard workflow for mRNA-seq library preparation that is described above is limited to examining populations of cells such as a pool of unicellular yeast growing the same condition, or a particular tissue from a mouse that is homogenized prior to extraction of RNA. The gene expression data that results from this type of library preparation therefore reflects an average across the cells that were used for extraction. Preserving heterogeneity in gene expression across cells in a dataset can provide valuable insights into important aspects of cell biology, such as cell type identification and function within a tissue, cellular differentiation, and drug resistance in cancer treatment [55]. In order to examine the variation in gene expression between individual cells in a population or tissue, methods for single-cell RNA sequencing (scRNA-seq) have been developed.

To perform scRNA-seq, individual cells must be separated prior to RNA isolation [6, 55]. A variety of methods have been used to isolate individual cells for scRNA-seq. These include diluting a sample to the level of a single cell and micromanipulation or microdissection to isolate individual cells from under a microscope. Other methods include fluorescence-activated cell sorting (FACS), in which fluorescently labeled cells are identified when they pass by a laser and are then separated from the rest of a population, and microfluidics to isolate individual cells in nanoliter-sized oil droplets.

Following single cell separation, RNA isolation and sequencing library preparation are performed. Sample processing for the individual cells that have been isolated occurs in steps very similar to the library preparation described previously, with two critical differences [6, 55]. First, each single-cell sample must be labeled with a unique barcode. These barcodes allow each sequencing read to be assigned to the cell from which it originated. Second, the amount of RNA yielded from a single cell is much smaller than in a typical bulk RNA-sequencing approach. First- and second-strand cDNA synthesis are performed on the RNA isolated from the individual cells (which can be done with polyA selection for analysis of mRNAs). However, to generate enough samples for sequencing, the cDNA is typically amplified via PCR. The amplification step can introduce substantial biases in the final sequencing library, which can be mitigated by the use of unique molecular identifiers (UMIs). UMIs are barcode sequences added to each library fragment during cDNA synthesis. After amplification, all library molecules that share a barcode can be traced back to a single starting RNA molecule, which allows for correction of any biases in the data during analysis.

This workflow has been further modified to allow for the preservation of spatial information for cells within a tissue [6, 55]. Isolation of a specific tissue section through laser capture microdissection (LCM) prior to single-cell separation can provide resolution on the location of cells with particular patterns of expression. Alternatively, mRNAs can be directly isolated from a tissue section by overlaying the tissue on a microarray chip with barcoded oligodT probes (Slide-seq). The barcodes are used to retain the spatial information for each RNA molecule in the tissue during data analysis.

The challenges in analyzing scRNA-seq data come first from associating the gene expression to a particular cell or cell type, then from filtering out the technical and biological variation that exists between samples and correcting for sample bias generated during library preparation [55]. Once these quality control steps are taken, the power of scRNA-seq allows for novel cell-subtypes to be defined within a sample, gene regulatory networks (genes that are coordinately regulated) to be elucidated, and cell fate specification to be determined.

Take Home Message

Exploration of RNA biology using genome-scale methods is now standard practice in molecular biology. The field of transcriptomics continues to expand and evolve, with new technologies and methods being developed at a rapid pace. Improvement to sequence quality and read length promise to continue to make these methods approachable for investigating a wide range of biological questions. Beyond the broad patterns of gene expression that can be readily assessed through sequencing, we can now delve deeper to investigate questions about transcriptome complexity, sample complexity, and a variety of questions in RNA biology beyond transcription.

References

1. Alwine JC, Kemp DJ, Stark GR. Method for detection of specific RNAs in agarose gels by transfer to diazobenzyloxymethyl-paper and hybridization with DNA probes. Proc Natl Acad Sci U S A. 1977;74(12):5350–4.
2. Wang AM. Quantitation of mRNA by the polymerase chain reaction. Proc Natl Acad Sci U S A. 1989;86(24):9717–21.
3. Adams M, Kelley J, Gocayne J, Dubnick M, Polymeropoulos M, Xiao H, et al. Complementary DNA sequencing: expressed sequence tags and human genome project. Science. 1991;252(5013): 1651–6.
4. Schena M, Shalon D, Davis RW, Brown PO. Quantitative monitoring of gene expression patterns with a complementary DNA microarray. Science. 1995;270(5235):467–70.
5. Mardis ER. The impact of next-generation sequencing technology on genetics. Trends Genet. 2008;24(3):133–41.

6. Stark R, Grzelak M, Hadfield J. RNA sequencing: the teenage years. Nat Rev Genet. 2019;20 (11):631–56.
7. Illumina Sequencing Platforms [Internet]. Illumina, Inc. [cited 2021 Jan 5]. https://www.illumina.com/systems/sequencing-platforms.html.
8. SEQC/MAQC-III Consortium. A comprehensive assessment of RNA-seq accuracy, reproducibility and information content by the Sequencing Quality Control Consortium. Nat Biotechnol. 2014;32(9):903–14.
9. Martin JA, Wang Z. Next-generation transcriptome assembly. Nat Rev Genet. 2011;12(10): 671–82.
10. Borodina T, Adjaye J, Sultan M. A strand-specific library preparation protocol for RNA sequencing. Methods Enzymol. 2011;500:79–98. https://linkinghub.elsevier.com/retrieve/pii/B9780123851185000050.
11. Conesa A, Madrigal P, Tarazona S, Gomez-Cabrero D, Cervera A, McPherson A, et al. A survey of best practices for RNA-seq data analysis. Genome Biol. 2016;17(1):13.
12. Sambrook J, Russell DW. Purification of RNA from cells and tissues by acid phenol-guanidium thiocyanate-chloroform extraction. CSH Protoc. 2006;2006(1):pdb.prot4045.
13. Vanecko S, Laskowski M Sr. Studies of the specificity of deoxyribonuclease I: III. Hydrolysis of chains carrying a monoesterified phosphate on carbon 5'. J Biol Chem. 1961;236(12):3312–6.
14. Sheng Q, Vickers K, Zhao S, Wang J, Samuels DC, Koues O, et al. Multi-perspective quality control of Illumina RNA sequencing data analysis. Brief Funct Genomics. 2017;16(4):194–204. https://doi.org/10.1093/bfgp/elw035.
15. Gallego Romero I, Pai AA, Tung J, Gilad Y. RNA-seq: impact of RNA degradation on transcript quantification. BMC Biol. 2014;12(1):42.
16. Evers DL, He J, Kim YH, Mason JT, O'Leary TJ. Paraffin embedding contributes to RNA aggregation, reduced RNA yield, and low RNA quality. J Mol Diagn. 2011;13(6):687–94.
17. Barbas CF III, Burton DR, Scott JK, Silverman GJ. Quantitation of DNA and RNA. CSH Protoc. 2007;2007:pdb.ip47.
18. Eddy SR. Non-coding RNA genes and the modern RNA world. Nat Rev Genet. 2001;2(12): 919–29.
19. Schroeder A, Mueller O, Stocker S, Salowsky R, Leiber M, Gassmann M, et al. The RIN: an RNA integrity number for assigning integrity values to RNA measurements. BMC Mol Biol. 2006;7 (1):3.
20. Dreyfus M, Régnier P. The poly(A) tail of mRNAs. Cell. 2002;111(5):611–3.
21. Kung JTY, Colognori D, Lee JT. Long noncoding RNAs: past, present, and future. Genetics. 2013;193(3):651–69.
22. Cabanski CR, Magrini V, Griffith M, Griffith OL, McGrath S, Zhang J, et al. cDNA hybrid capture improves transcriptome analysis on low-input and archived samples. J Mol Diagn. 2014;16(4):440–51.
23. Curion F, Handel AE, Attar M, Gallone G, Bowden R, Cader MZ, et al. Targeted RNA sequencing enhances gene expression profiling of ultra-low input samples. RNA Biol. 2020;17 (12):1741–53.
24. Head SR, Komori HK, LaMere SA, Whisenant T, Van Nieuwerburgh F, Salomon DR, et al. Library construction for next-generation sequencing: overviews and challenges. Biotechniques. 2014;56(2):61–4. https://www.future-science.com/doi/10.2144/000114133.
25. Hrdlickova R, Toloue M, Tian B. RNA-Seq methods for transcriptome analysis: RNA-Seq. WIREs RNA. 2017;8(1):e1364.
26. Pelechano V, Steinmetz LM. Gene regulation by antisense transcription. Nat Rev Genet. 2013;14 (12):880–93.

27. Dominic Mills J, Kawahara Y, Janitz M. Strand-specific RNA-Seq provides greater resolution of transcriptome profiling. Curr Genomics. 2013;14(3):173–81.
28. Illumina, Inc. An introduction to next-generation sequencing technology. San Diego: Illumina; 2017.
29. Chen F, Dong M, Ge M, Zhu L, Ren L, Liu G, et al. The history and advances of reversible terminators used in new generations of sequencing technology. Genomics Proteomics Bioinform. 2013;11(1):34–40.
30. Illumina, Inc. 2-channel SBS Technology. San Diego: Illumina; 2021. https://www.illumina.com/science/technology/next-generation-sequencing/sequencing-technology/2-channel-sbs.html.
31. Illumina, Inc. Illumina CMOS chip and one-channel SBS chemistry, vol. 4. San Diego: Illumina; 2018.
32. Illumina, Inc. Paired-end vs. single-read sequencing technology. San Diego: Illumina; 2021. https://www.illumina.com/science/technology/next-generation-sequencing/plan-experiments/paired-end-vs-single-read.html.
33. Kukurba KR, Montgomery SB. RNA sequencing and analysis. Cold Spring Harb Protoc. 2015;2015(11):pdb.top084970.
34. Moreton J, Izquierdo A, Emes RD. Assembly, assessment, and availability of de novo generated eukaryotic transcriptomes. Front Genet. 2016;6:361. http://journal.frontiersin.org/Article/10.3389/fgene.2015.00361/abstract.
35. Pertea M, Kim D, Pertea GM, Leek JT, Salzberg SL. Transcript-level expression analysis of RNA-seq experiments with HISAT, StringTie and Ballgown. Nat Protoc. 2016;11(9):1650–67.
36. Morlan JD, Qu K, Sinicropi DV. Selective depletion of rRNA enables whole transcriptome profiling of archival fixed tissue. PLoS One. 2012;7(8):e42882.
37. Atkinson SR, Marguerat S, Bähler J. Exploring long non-coding RNAs through sequencing. Semin Cell Dev Biol. 2012;23(2):200–5.
38. Li J, Liu C. Coding or noncoding, the converging concepts of RNAs. Front Genet. 2019;22(10):496.
39. Choudhuri S. Small noncoding RNAs: biogenesis, function, and emerging significance in toxicology. J Biochem Mol Toxicol. 2010;24(3):195–216.
40. Liu Q, Ding C, Lang X, Guo G, Chen J, Su X. Small noncoding RNA discovery and profiling with sRNAtools based on high-throughput sequencing. Brief Bioinform. 2019;22(1):463–73.
41. Zhang P, Wu W, Chen Q, Chen M. Non-coding RNAs and their integrated networks. J Integr Bioinform. 2019;16(3):20190027. https://www.degruyter.com/view/journals/jib/16/3/article-20190027.xml.
42. Dard-Dascot C, Naquin D, d'Aubenton-Carafa Y, Alix K, Thermes C, van Dijk E. Systematic comparison of small RNA library preparation protocols for next-generation sequencing. BMC Genomics. 2018;19(1):118.
43. Calviello L, Ohler U. Beyond read-counts: Ribo-seq data analysis to understand the functions of the transcriptome. Trends Genet. 2017;33(10):728–44.
44. Wissink EM, Vihervaara A, Tippens ND, Lis JT. Nascent RNA analyses: tracking transcription and its regulation. Nat Rev Genet. 2019;20(12):705–23.
45. Ganser LR, Kelly ML, Herschlag D, Al-Hashimi HM. The roles of structural dynamics in the cellular functions of RNAs. Nat Rev Mol Cell Biol. 2019;20(8):474–89.
46. Strobel EJ, Yu AM, Lucks JB. High-throughput determination of RNA structures. Nat Rev Genet. 2018;19(10):615–34.
47. Wang B, Kumar V, Olson A, Ware D. Reviving the transcriptome studies: an insight into the emergence of single-molecule transcriptome sequencing. Front Genet. 2019;26(10):384.
48. Rhoads A, Au KF. PacBio sequencing and its applications. Genomics Proteomics Bioinform. 2015;13(5):278–89.

49. Kono N, Arakawa K. Nanopore sequencing: review of potential applications in functional genomics. Develop Growth Differ. 2019;61(5):316–26.
50. Piovesan A, Antonaros F, Vitale L, Strippoli P, Pelleri MC, Caracausi M. Human protein-coding genes and gene feature statistics in 2019. BMC Res Notes. 2019;12(1):315.
51. Adams MD, Celniker SE, Holt RA, Evans CA, Gocayne JD, Amanatides PG, et al. The genome sequence of Drosophila melanogaster. Science. 2000;287(5461):2185–95.
52. Miura F, Kawaguchi N, Yoshida M, Uematsu C, Kito K, Sakaki Y, et al. Absolute quantification of the budding yeast transcriptome by means of competitive PCR between genomic and complementary DNAs. BMC Genomics. 2008;29(9):574.
53. Schwartz S, Motorin Y. Next-generation sequencing technologies for detection of modified nucleotides in RNAs. RNA Biol. 2017;14(9):1124–37.
54. Xu L, Seki M. Recent advances in the detection of base modifications using the nanopore sequencer. J Hum Genet. 2020;65(1):25–33.
55. Hwang B, Lee JH, Bang D. Single-cell RNA sequencing technologies and bioinformatics pipelines. Exp Mol Med. 2018;50(8):96.

The mRNA and the New Vaccines

6

Anjali Desai and Neena Grover

Contents

Keywords

mRNA structures · mRNA lifecycle · RNA genome mapping · mRNA vaccine · Vaccine technology RNA

A. Desai · N. Grover (✉)
Department of Chemistry and Biochemistry, Colorado College, Colorado Springs, CO, USA
e-mail: Anjali.desai@coloradocollege.edu; ngrover@ColoradoCollege.edu

© Springer Nature Switzerland AG 2022
N. Grover (ed.), *Fundamentals of RNA Structure and Function*, Learning Materials in Biosciences, https://doi.org/10.1007/978-3-030-90214-8_6

What You Will Learn

The messenger RNA molecules provide the order in which the amino acids are linked together, via a register of three-nucleotide sequences (codons). The structures in mRNA contribute to its many functions including, spatial and temporal regulation of translation and its own degradation. This chapter is a brief introduction to mRNA. Many associated topics are covered in other chapters.

We have entered an era of RNA-based technologies. Vaccines based on mRNA are quick to design and have proven effective so far. Vaccine development using mRNA has been tested for at least a decade, long before the current coronavirus-caused pandemic. RNA-based therapies are likely here to stay.

Learning Objectives

After reading this chapter, the students should be able to:

- Define the common structural features of mRNA.
- Discuss the role of mRNA structures beyond coding for proteins.
- Explain the life cycle of mRNA and its role in regulation of proteins.
- Explain the principles of mRNA-based vaccine technology.

6.1   Introduction

A small percent (1–2%) of human genome codes for proteins [1]. Heteronuclear RNA (hnRNA) or pre-mRNA is the complement of the coding regions in the DNA (genes). The hnRNA are processed into the messenger RNA (mRNA) through RNA processing (splicing, editing and modifications) (Chap. 4, Spliceosome) [1–12]. A single hnRNA can produce multiple mature mRNA isoforms; each mRNA produces a different protein.

Approximately 90% of human genes undergo splicing and alternate splicing. Every tissue in the body is characterized by unique splicing events, with heart, brain, and skeletal muscles showing the most highly conserved and tissue-specific alternate splicing patterns [5, 6]. For example, the gene for a protein Titin, which is important for heart contractions, produce several different length isoforms; each protein has a unique function and is developmentally regulated. In eukaryotic organisms or viruses that undergo splicing, the DNA information is nonlinear and produces multiple mRNA isoforms, thus less than 30,000 genes produce between 100,000 and 400,000 proteins.

The mRNA undergoes several processing steps in the nucleus before it moves out to the cytoplasm or is localized to different cellular organelles [12]. Every step in the way has a quality control process to ensure that the correct mRNA is being sent to the ribosomes; all

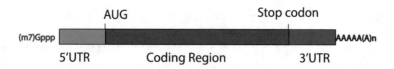

Fig. 6.1 Key features of mRNA. The mRNA has a 5′ cap (m⁷G cap in eukaryotes and a 5′ triphosphate cap in bacteria), a 5′ UTR, a coding region defined by a start and stop codon, a 3′ UTR, and often a poly(A) tail

incorrect mRNA are marked for degradation. Translation of defective mRNA lead to defective proteins, which cause disease.

6.2   Key Features in mRNA

The key features of mRNA include: a 5′ untranslated region (5′ UTR), a coding region (CDS), and a 3′ untranslated region (3′ UTR). The eukaryotic mRNA contain: a 5′ cap and a 3′ polyadenylated tail (Fig. 6.1).

Secondary and tertiary structural features of mRNA contain information regarding its packaging, splicing, and translation speed [1–12]. Genome-wide studies are elucidating the role of mRNA structures beyond the codons [5–10]. The spatial and temporal expression of mRNA, along with their sequestration into P-bodies shine a light on mRNA-based regulatory mechanisms.

6.3   Bacterial Messenger RNA

In bacteria, the RNA is complement of DNA. The process of transcription and translation occur simultaneously due to a lack of a nucleus and intermediary processing steps (Chap. 10, Transcription) [3, 4]. Bacterial mRNA transcripts consist of a 5′ triphosphate cap, a 5′ UTR with a ribosome-binding site (RBS) (also called the Shine-Dalgarno, SD, sequence), a coding region, a 3′ UTR, and sometimes a 3′ poly(A) tail. These structural features are key points of regulation for expression of the genetic code.

RNA Polymerase (RNAP) catalyzes bacterial transcription after binding to a promoter on the DNA template. The transcripts can code for more than one protein and are thus said to be polycistronic. A single promoter controls transcription of a polycistronic transcript; each protein-coding sequence has its own start and stop codon.

As soon as an mRNA transcript is produced from DNA, *cis*-acting sequences that are a part of the nascent mRNA chain, recruit ribosomes for translation. The ribosome-binding site (RBS) has a consensus sequence of AGGAGG which lies about 8–12 nucleotides upstream of the start codon (AUG). Ribosome base pairs with the RBS to begin translation.

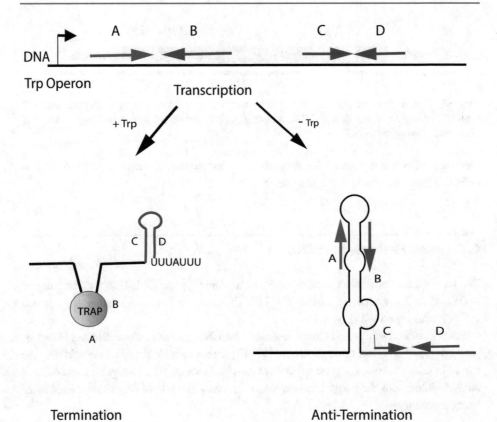

Termination   Anti-Termination

Fig. 6.2 Attenuation of transcription. In the operon for tryptophan (Trp), the triplet repeats in the A and B region, GAG or UAG, bind to 11 subunits of TRAP protein (shown as a circle) that causes C and D regions to form a hairpin structure that causes termination. In the absence of tryptophan, A and B regions form a stem-loop structure using the repeat sequences that preclude the formation of the termination sequence, thus forming anti-termination hairpin structures and allowing the transcription of downstream genes to continue [13]

Bacterial Regulation of Transcription and Translation Bacteria regulate transcription and translation through *cis*- and *trans*-acting factors. *Cis*-acting factors are part of the nascent mRNA chain and *trans*-acting factors are additional molecules that bind to the RNA. Transcription attenuation is often seen in operons in bacteria and archaea [13]. It is a provisional stop sign that can form in the RNA leader sequence in order to stop the transcription. This is a $5'$-*cis*-acting regulatory mechanisms—that is, the structures of RNA can cause stalling of the ribosome in response to the cell's needs. Attenuation can result in early transcription termination or transcription of additional downstream genes of an operon by forming alternative hairpin (anti-terminator) structure during transcription (Fig. 6.2).

An operon contains many genes under the control of one promoter. The terminator, anti-terminator, or anti-anti-terminator structures form in a polycistronic mRNA to control the length of the mRNA transcript produced, thus, controlling the specific genes that are transcribed [13]. Following the terminator sequence is often a string of U residues, that promote RNAP dissociation, thus, ending transcription [13]. Besides secondary structures, the RBS is another *cis*-acting regulator. The RBS promotes translation by binding the ribosomes on to the mRNA; the sequences that contain consensus SD are more efficiently translated by the ribosome [3] (Chap. 10, Transcription).

Riboswitches are *cis*-acting regulatory sequences found in the 5′ UTR of mRNA that bind to small molecules [3, 7, 8, 14]. Riboswitches can stabilize secondary structures upstream of a terminator attenuation sequence, thus terminating transcription. They can form secondary structures that sequester the RBS to inhibit translation. Different metabolic signals control formation and dissolution of the riboswitch secondary structures. A summary of riboswitch-based regulation of mRNA is shown in Fig. 6.3 [14] (Chap. 7, Riboswitch).

Degradation The degradation of mRNA allows a cell to adjust its gene expression, recycle RNA components, respond to extracellular signals, and eliminate aberrant, viral, or toxic RNA. In bacteria, degradation generally proceeds through the use of endonucleases that cleave in the middle of a transcript. Endonucleases like RNase III recognize and cleave hairpin structures. Some bacterial mRNA have poly(A) tails that marks these for degradation. Often, bacterial degradation enzymes are organized into a complex called degradasome. These complexes contain helicases, exonucleases, and endonucleases which degrade a majority of bacterial mRNA [15].

6.4 Eukaryotic Messenger RNA

In eukaryotes, mRNA is transcribed and processed in the nucleus before being exported to the cytoplasm for translation (Fig. 6.4). Eukaryotic mRNA contains a 5′ 7-methyl guanosine cap and a poly(A) tail in addition to the features present in the bacterial mRNA—a 5′ UTR, a coding region, and a 3′ UTR.

Transcription Transcription takes place in the nucleus. With the help of transcription factors, RNA Polymerase II (RNA Pol II) begins de novo transcription of pre-mRNA using a promoter region on the DNA template. RNA Pol II contains a unique carboxyl-terminal domain (CTD) that acts as a recruitment scaffold for processing factors [3, 16, 17]. The recruitment of mRNA processing machinery occurs co-transcriptionally along with the enzymatic modification of mRNA [18]. The CTD can also help recruit proteins that will eventually allow the mRNA to exit the nucleus [19].

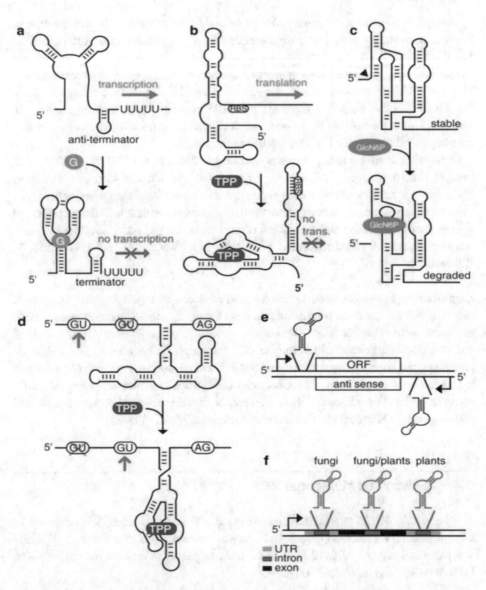

Fig. 6.3 Riboswitch-based regulation of mRNA. (**a**) A guanine-sensing riboswitch binds guanine to form a terminator stem and causes transcription to terminate. (**b**) The TPP-riboswitch causes translational regulation by TPP. In the absence of TPP, downstream genes are translated as RBS is available. In the presence of TPP, an alternate conformation blocks the RBS and halts translation. (**c**) The GlcN6P riboswitch is cut in the presence of high amounts of GlcN6P, inhibiting translation of mRNA. (**d**) Alternate splicing by TPP-riboswitch. In the absence of TPP, the base pairing allows only the first 5′ SS to be available. The nucleotides binding the second splice site only become available when TPP binds, thus allowing alternate splicing. (**e**) Eubacterial riboswitch is present in the 5′ UTR and directly controls the expression of downstream genes. Some control expression of an antisense RNA to regulate protein expression from a different mRNA (indirect control). (**f**) Eukaryotic riboswitches are found in 5′ UTR, coding regions, and 3′ UTR. (Figure from [14])

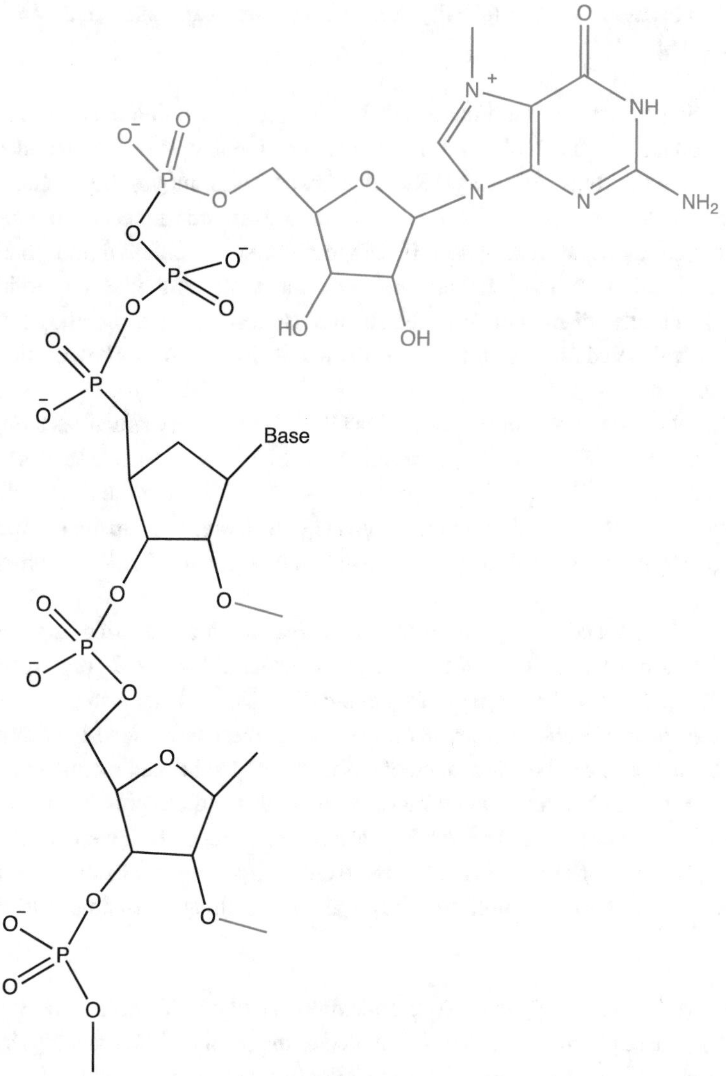

Fig. 6.4 The 5′ cap. The N7-methyl guanosine cap (red) has 5′ to 5′ connection and a methyl group on N7. The second and third sugars may also have a 2′-*O* methyl

5′ UTR Untranslated regions (UTRs) flank the coding sequence on both the 5′ and 3′ end of mRNA. These regions contain many sequences, including the 5′ cap and 3′ poly(A) tail, which determine a mRNA's function and stability. The 5′ cap protects the mRNA from 5′ to 3′ exonucleases and phosphatases. A cap-binding complex (CBC) attaches to the 5′ cap before splicing. This complex plays a role in splicing, polyadenylation of the 3′ end, nuclear export, and degradation. In the cytoplasm, the cap is crucial for translation. A translation initiation factor (eIF4E) recognizes the cap and aids in ribosome recruitment [16].

The addition of a 5′ cap onto the 5′ UTR is the first processing event and takes place co-transcriptionally. After RNA Pol II has transcribed the first 25–30 nucleotides, capping enzymes bind to the RNA Pol II. First RNA triphosphatase removes the Υ-phosphate from the first nucleotide triphosphate. Then guanylyl transferase adds guanosine monophosphate (GMP) from a guanosine triphosphate (GTP) to the first nucleotide diphosphate (rNDP). This forms a unique 5′-to-5′ linkage between the GMP and first nucleotide on the pre-mRNA. Finally, 7-methyl transferase methylates guanine at the N7 position (Fig. 6.4). After the addition of the 5′ cap, RNA Pol II is no longer held at the promoter and elongation begins [16].

The 5′ UTR of eukaryotic mRNA contains GC-rich secondary structures along with the Kozak sequence which facilitates translation. The Kozak sequence has a consensus sequence: GCCGCCRCCAUGG (R = purines), with AUG being the start codon. The 5′ UTR contains the internal ribosome entry sites (IRES) along with upstream open reading frames (uORFs) to allow for differential translation of a given mRNA transcript [3].

3′ UTR The 3′ UTR of eukaryotic mRNA contains structural features important for the mRNA's function. Examples include, AU-rich elements (AREs) and poly(A) tails. AREs affect mRNA stability and mark them for degradation. The 3′ UTR also contains information for localization of RNA to different organelles and these have cleverly been labeled zip codes [12]. Zip codes can be a few nucleotides or >1 kb in size and recruit a particular set of proteins. Each mRNA can contain multiple zip codes and the proteins associated with the zip codes associate with the cytoskeletal motors to regulate transport. The level of structure in these regions of RNA may provide clues regarding localization and regulation of mRNA. It is important to note that localization signals are also found in the coding region [8].

All eukaryotic mRNAs contain 50–250 adenines called poly(A) tails at the 3′ end. In the final step of transcription, a poly(A) tail is added to the 3′ end of the transcript by cleaving the transcript. In mammals, the cleavage site lies between the consensus sequence AAUAAA and a U/GU-rich region. Cleavage/polyadenylation specific factor (CPSF) recognizes the AAUAAA sequence, cleaves the transcript and recruits a poly (A) polymerase (PAP). The PAP adds a poly(A) tail to the 3′-OH group of the exposed nucleotide. Then RNA Pol II dissociates from the rest of the transcript, allowing the cleaved piece to be degraded. Different cleaving sites for CPSF in the 3′ UTR allow for alternative polyadenylation of mRNA transcripts. These transcripts are isoforms of each other that can encode completely different proteins, have different stabilities, localization signals, and tissue specificities [16].

The poly(A) tail protects mRNA from 3′ to 5′ exonucleases, determines its stability, and helps direct its localization to different regions of a cell. Additionally, a poly(A)-binding protein (PAB) bound to the poly(A) tail can interact with the translational machinery to circularize mRNA to offer further stabilization and facilitate translation [16].

6.5   Viral RNA

Many viruses have a RNA-based genome [20–23]. The RNA genome itself is often the mRNA. For example, coronavirus SARS-CoV-2 has a very structured RNA-based genome that is ready to be translated upon entry into the host cells [20, 21] (Fig. 6.5). In retroviruses, like HIV-1, reverse transcriptase makes a DNA copy of its RNA genome that is then integrated into the host DNA [22]. Retroviruses encode the reverse transcriptase and integrase enzymes to accomplish these tasks. The viral genomes are small and contain only the information necessary to make essential proteins including those needed for replication and modulating host immune response. (Viroids are ~250–450 nucleotide, circular, single-stranded RNA that don't code for proteins but can still be pathogenic.)

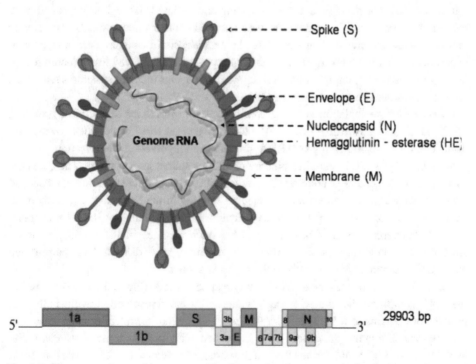

Fig. 6.5 The coronavirus and its genome. The Coronavirus is composed of a single-stranded RNA genome. It is covered with phosphorylated nucleocapsid (N) proteins. The nucleocapsid is buried within a phospholipid bilayer that contains trimeric glycoprotein spike protein (S), membrane protein (M), hemagglutinin-esterase (HE) and the envelope protein (E). The positive single-stranded RNA-based genome is 29.9 kb mRNA. (Figure is from [21] under open CC 4.0 and has not been altered)

6.6   Genome Mapping of RNA

The structures of genomes of HIV-1, yeast, and several plant and mammals have been mapped in vitro and in vivo using enzymatic and chemical probing methods and by genome-wide studies [6–10, 24]. A study of RNA structures—RNA structuromes—reveals regulatory effects of RNA structures on mRNA polyadenylation, splicing, translation, and turnover. Modifications of mRNA, in particular methylation and pseudo-uridylation, alter RNA structure, stability, and function. The genome mapping experiments highlight the plasticity of RNA structures and the importance of experimental conditions used to study RNA. Under high ionic conditions typical for in vitro experiments, different RNA structures were stabilized as compared to those seen in vivo experiments [7–9]. A key difference might be that RNA under in vivo conditions are bound up by cellular factors that could stabilize alternate structures; nonetheless, it is advisable to consider mRNA structures in vitro as potential structures that might form under in vivo conditions.

The structure in mRNA shows a triplet repeating pattern in the coding regions that is absent from the UTR regions and is thought to minimize the ribosome slippage during translation [10]. The average reactivity of coding regions and UTR regions appears to vary between organisms and is expected to vary between cellular compartments. The coding region is more structured than the UTRs in *Arabidopsis* in both in vitro and in vivo experiments. Nuclear transcripts, predominantly pre-mRNA, showed less structured coding regions than UTR, implying that RNA processing events increase structures within the coding regions and decreases them in UTRs.

Alternate polyadenylation is linked to differences in reactivity of chemical probes at 15–22 nucleotides upstream (-22 to -15 positions) and -1 to $+5$ nucleotides downstream of the cleavage site, indicating that RNA structures are involved in site selection.

Splicing of pre-mRNA requires binding of various spliceosomal factors. Thus, it is not surprising that the choice of splicing sites, along with splicing efficiency, are sequence- and structure-dependent. Thermodynamic structural stability of mRNA is also correlated with its decay rates, suggesting a role of RNA structures in the unfolding of RNA for degradation. Modification sites in RNA also correlate with changes in the stability of RNA structures and their associated functions. For example, m^6A disrupts base pairing and was found to direct splicing factors to the opposite strands.

The HIV-1 genome was the first to be fully mapped in vivo (Fig. 6.6). The structures in the RNA controlled the rate of protein synthesis by the ribosome, allowing individual proteins or domains to fold. The RNA was more structured in protein-domain junctions. The regions of highly structured RNA correlated with protein tertiary structure. For example, protein loops are derived from highly structured elements in RNA. The frameshift region formed a three-helix junction and not the predicted stem-loop adjacent to the slippery sequences containing multiple uracil.

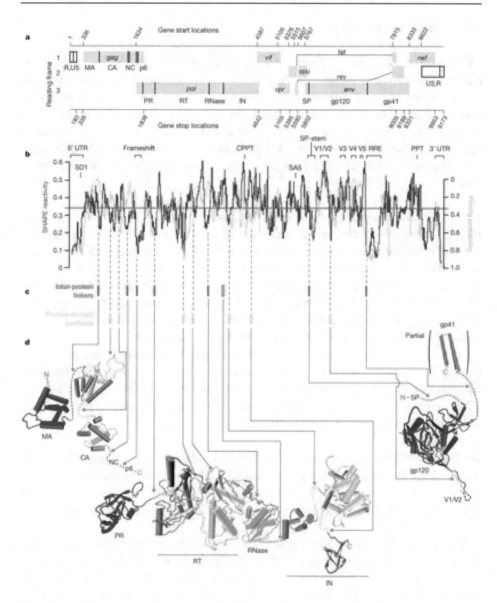

Fig. 6.6 HIV-1 genome mapping. (**a**) The genome organization of HIV-1. (**b**) The in vivo chemical probing of RNA structures established a link between RNA structure and its reactivity. Specific regions of RNA are identified above the reactivity plot (*SA* splice acceptor, *SD* splice donor, *PPT* polypurine tract, *cPPT* central polypurine tract, *V* variable regions, *RRE* Rev response element). The reactivity of RNA (dark blue) and the pairing probabilities (light blue) are plotted on the *y*-axis. (**c**) Interdomain linker (green) and protein domains (yellow) are marked. (**d**) The folded protein structures (blue, red, light magenta, purple, and red) are linked to the RNA accessibility to probes. (Figure from [24])

6.7   RNA Processing

Splicing In eukaryotes, a single RNA polymerase II generates the pre-mRNA. Spliceosomes assemble on pre-mRNA as it is being transcribed [18]. Splicing removes the intervening sequences (introns) in pre-mRNA and ligates expressed sequences (exons) together (Chap. 4, Spliceosome). The co-transcriptional events 5′-capping, RNA modifications, 3′-end processing all assist in the choice of introns and exons. In mammals and yeast, splicing involves two transesterification reactions. The structures in RNA play a role in splice site selection and assembly of the spliceosome and influence the rate of transcription.

Various *cis*- and *trans*-acting regulatory factors influence splicing. In mammals, exons are recognized by *trans*-factors, such as serine/arginine-rich (SR) proteins that enhance splicing. Different expression levels and binding patterns of these *cis*- and *trans*-acting regulatory factors allow multiple mature mRNA isoforms to form (alternative splicing). Splicing and alternative splicing create different functional proteins from the same pre-mRNA. Nearly 90% of all mammalian genes undergo alternate splicing and contribute to the diversity of proteins expressed in the cell. Splicing plays a key role in human immune response and allows the RNA to respond to the environmental cues [17].

After the production of mRNA in the nucleus, protein complexes bind mRNA for export. The SR proteins are removed to reveal binding sites for messenger ribonucleoprotein (mRNP). Export factors are recruited to the transcript in a splice-dependent manner. Many of the proteins that bind to mRNA at this stage are carried with it to the cytoplasm and have important role in its fate [16] (Fig. 6.7).

Editing Pre-mRNA goes through editing during and after transcription. Editing includes processes that alter or add nucleotides to the RNA [25]. Two common types of editing are deamination of nucleotides (adenine to inosine; cytosine to uracils) or insertion/deletion of nucleotides. Multi-protein complexes called editosomes catalyze these reactions [25].

Changes in codon usage due to editing can changes protein expression as is another mechanism of adding to diversity. When cytosine is deaminated to make uridine, it converts a CAA codon into a UAA stop codon, resulting in a truncated protein. This truncation can direct the protein to a specific tissue. When adenosine is deaminated to make inosine, it is read as guanosine. Additionally, adenosine to inosine deamination changes the base pairing interactions within RNA, altering its secondary structures [25].

In some protozoa, uridine nucleotides are added to and deleted from transcripts in the mitochondria. These changes can add start, stop, and functional amino acid codons to vary the proteins encoded by any given sequence. Different stages of development control these insertions and deletions [26].

The introduction of an AUG start codon in an already transcribed mature mRNA transcript could result in a new protein without de novo synthesis of a new mRNA.

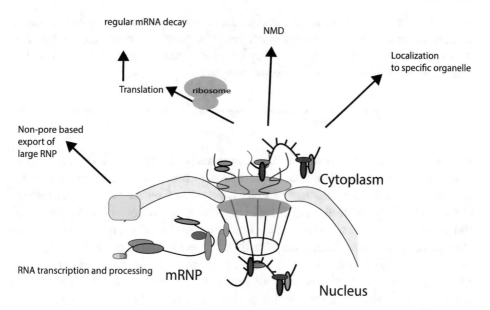

regular mRNA decay

NMD

Localization
to specific organelle

Translation ribosome

Non-pore based
export of
large RNP

Cytoplasm

RNA transcription and processing mRNP

Nucleus

Fig. 6.7 The life cycle of mRNA. Once hnRNA is converted to mRNA, it is transported out of the nucleus. The mRNA exits the nucleus as a RNA-protein complex (mRNP). Once in the nucleus, RNA is immediately translated, localized to appropriate organelles, or degraded

Thus, editing adds a large amount of diversity from a single gene and allows the genetic code to be relatively small [25, 26].

Quality Control The quality of RNA made in the nucleus goes through many checks to ensure that the correct mRNA are being transported out of the nucleus [27–31]. Aberrant mRNA are degraded in the nucleus and do not accumulate [28]. The nuclear exosome is a ten-protein complex that monitors the integrity of mRNA transcripts in the nucleus. It also degrades mRNA transcripts that accumulate within the nucleus through a process termed DRN (decay of RNA in the nucleus) using $3'$ to $5'$ riboexonuclease activity [27]. Those mRNA that are correctly processed are likely to be protected by the proteins that are bound to it.

Quality control also occurs at the nuclear pore called Mlp/Tpr surveillance. Mlr is myosin like protein in budding yeast; Tpr is a translocated promoter region in vertebrates [29, 30]. These proteins are located in the inner basket of the nuclear pore complex (NPC). They make contacts with mRNP as it moves toward export from the nucleus. In one model, Mlp interacts with splicing factor 1 and retains the mRNA until processing is completed. Splicing occurs while RNA is tethered to the inner face of the nuclear pore complex and remains there until splicing is completed. In another model, Mlp proteins bind to fully processed transcripts to concentrate them at the nuclear pore for export. Mlp2 binds to nuclear export factor Yra1p.

Two translation-dependent processes also control mRNA degradation in the cytoplasm: the nonsense mediated decay (NMD) which recognizes premature termination codons and other imperfections and the non-stop decay which recognizes transcripts that lack a stop codon. Exosomes also degrade RNA in the cytoplasm.

6.8   Life Cycle of mRNA

RNA is transcribed and processed from hnRNA to mRNA in the nucleus. Once it is made, it undergoes further surveillance at the nuclear pore before being exported out of the nucleus. In the cytoplasm, the fate of RNA is determined by the proteins that bind to it. The mRNA may get localized to different organelles, be further checked for defects by NMD, non-stop decay and exonucleases. Ribosomes are recruited on to mRNA as it is exported out of the pore. Translation of mRNA and its lifetime in the cell are important points of regulation in producing specific amounts of proteins. The structures found in the mRNA serve multiple functions, the details of which are only just being uncovered via in vivo structure probing, genome analyses, and large-scale comparative analyses.

Nuclear Export The pre-mRNA is converted to mRNA in the nucleus and must be exported to the cytoplasm through the nuclear pore complex (NPC) [32–34]. The NPC is embedded in the nuclear membrane. Packaging of mRNA for export is coupled to transcription [33]. Two yeast mRNA export proteins, Npl3 and Yra1, associate with RNA polymerase II during transcription and are likely to remain bound to the mRNA for its export. Npl3 is an abundant nuclear export protein that is likely to package mRNA for export; it leaves the nucleus as part of hnRNPs and is removed after export and returns to the nucleus. Yra1 marks the mRNA for completion of RNA processing. Defects in mRNA processing are seen to occur when mRNA export process is defective. Once at the NPC, export factors bind to NPC channel proteins in a Ran-dependent process to allow the mRNA to pass into the cytoplasm in a $5'$ to $3'$ manner. In the cytoplasm, certain nuclear proteins are replaced with cytoplasmic proteins to prevent mRNA from reentering the nucleus [32].

Large RNA–protein complexes that do not fit into the nuclear pore, such as those involved in localization of certain neuronal mRNA, may use a budding method for exiting the nucleus (nuclear envelope budding) [34].

Localization After export into the cytoplasm, some of the mRNA are localized to different regions of the cell before they can be translated [31]. Differences in mRNA localization allows for local production of proteins for rapid and local response to stimuli along with differential delivery of proteins to the organelles. For example, the locally synthesized proteins serve many different functions in developing and mature axons to rapidly respond to extracellular stimuli and to different physiological states. The regulation of cohorts of functionally related mRNA (RNA regulons) drive axon growth and guidance, injury response, survival, and axonal mitochondrial functions [35].

Translation Repression, P-bodies and Stress Granules Translation of mRNA often occur in the cytoplasm. Continuous translation of all mRNA transcripts can be energetically expensive. When mRNAs are not needed, their translation can be repressed by their sequestration into granules. P-bodies are granules that sequester specific non-translating mRNA [31]. P-bodies contain decapping enzymes, exonucleases, and other degradation machinery. P-bodies can begin mRNA degradation or quarantine transcripts until these are required for translation [31]. Stress granules form when stress leads to inhibition of translation initiation. These sequester mRNA along with transcription initiation factors, small ribosomal subunits, and poly(A) binding proteins. This protects mRNAs from degradation. Once the stressors have been relieved, stress granules release the mRNA transcripts along with initiation factors, ready for translation [31].

Mature mRNA is ready to be used by the ribosome to make proteins. If a mRNA does not need to be localized, translation initiation can begin as soon as the 5' cap-binding complex (CBC) protrudes into the cytoplasm from the NPC. This begins by replacement of the CBC by a translation initiation factor. Ribosomes are recruited to the mRNA and translation can begin [32].

Many different features of mRNA regulate the rate of translation. The ribosome recognizes the Kozak sequence in the 5' UTR to help identify the first AUG codon. The closer the contextual sequence around a particular AUG is to the consensus Kozak sequence, the more likely a ribosome is to bind to and transcribe the mRNA transcript. Some secondary structures in the 5' UTR can inhibit ribosomal scanning of the transcript or enhance ribosomal attachment to IRES (internal ribosome entry site). If several open reading frames are detected, uORFs compete with the main open reading frame for translation initiation [3]. Different cofactors fine-tune these regulatory factors to meet a cell's needs.

Degradation as a means of regulation   At the end of mRNA's life, it must be degraded to allow a cell to adjust the gene expression to its needs, recycle RNA components, respond to extracellular signals, and eliminate aberrant, viral, or toxic RNA. The rate of mRNA degradation is a mode of regulation [3, 15, 32–36]. Some of the key steps are below.

In eukaryotes, 5' cap and 3' poly(A) tail interact with the cytoplasmic elF4E and poly (A)-binding proteins (PABP), respectively, in order to protect the RNA from degradation and enhance translation initiation. To initiate degradation either RNA has to be cleaved by endonucleases or the 5' cap/3' poly(A) tail structures must be compromised. The bulk of mRNA degradation begins with deadenylation of the poly(A) tail. After deadenylation, decapping enzymes can remove the 5' cap and begin 5' to 3' degradation by XRN1 exoribonuclease, or 3' to 5' exonucleases (exosomes) can begin degradation at the 3' end

[36] (Fig. 6.8). The exosome is 10–12 subunit complex that plays a role in 3′-end processing of mRNA and ncRNA. The degrading enzymes are assisted by RNA helicases that unwind the secondary structures of RNA [15]. Surveillance machinery in the nucleus scans transcripts for mistakes in capping, splicing, polyadenylation, and export. If mutations or improper processing are detected, the transcript is marked for degradation.

Rates of degradation are affected by many *cis*-acting factors. A longer poly(A) tail indicates a more stable mRNA transcript. AREs in the 3′ UTR destabilize the mRNA structure and mark it for rapid degradation. Iron response elements in the 5′ and 3′ UTR bind iron response proteins, and depending on the level of iron in the cell, mark the transcript for degradation or stabilize it [3].

Other means of rapid degradation involve *trans*-acting elements. MicroRNA are sequestered in P-bodies and have sequences that are complementary to the mRNA (Chap. 8, ncRNA). Binding of miRNA to mRNA marks it for rapid degradation.

Nonsense mediated decay (NMD) is used when a premature termination codon is detected in a transcript; this degradation can also be carried out in P-bodies [31].

6.9   The Development of mRNA Vaccines

Coronavirus pandemic has brought in a new era of mRNA vaccines. To understand the speed of mRNA vaccines development, we will briefly discuss our evolving understanding of the immune system, the role lipids in RNA delivery, and decades of research on the coronaviridae family of viruses.

Viral Life Cycle Viruses are small (~0.1 μm diameter for coronavirus) compared to a eukaryotic cell (10–100 μm diameter). They have a DNA- or RNA-based genome and a small set of proteins that are virus-specific (i.e., not found in the host cell). Viruses utilize the host cell's enzymes, building blocks (nucleotides, amino acids), and cofactors (ex: magnesium) to replicate. This requires making multiple copies of their RNA (genome) and proteins.

For RNA viruses, the viral genome copies look like the host cell's mRNA with a 5′ m^{7}G cap and 3′ poly(A) tail. mRNA are translated by cellular ribosomes to make viral proteins. When multiple copies of viral genome and proteins become available, these assemble into new virus particles. The new viruses leave the cell (viral budding).

Cell surface proteins on the new virus interact with a new host cell to gain entry. For example, the trimeric spike protein (S), which forms the crown of coronavirus, binds to the angiotensin-converting enzyme (ACE2) on the outside of human cell to transfer its genome and proteins inside the cell [20, 21].

Immune Response to Proteins Before we understood biology as we know it today, we understood the idea of immunity. Countries in Asia had methods to train the immune

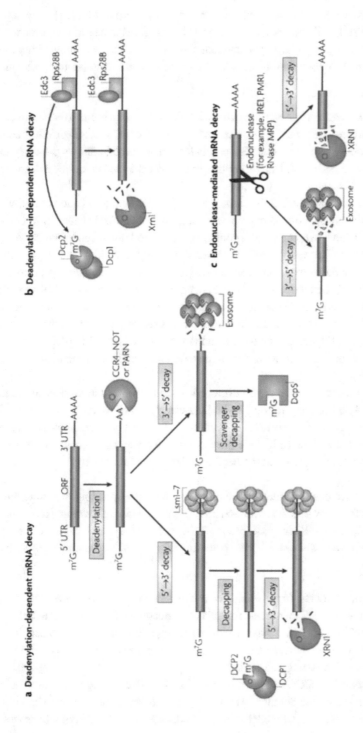

Fig. 6.8 Decay of mRNA. (**a**) mRNA are deadenylated by CCR4-NOT or PARN. The Lsm1–7 complex binds to the 3′ end of the mRNA to induce DCP1-DCP2 dependent decapping. This leaves RNA susceptible to XRN1 nuclease. Alternatively, RNA is degraded in the 3′ to 5′ direction after deadenylation. (**b**) In yeast, the deadenylation-independent pathway for mRNA decay first uses the decapping nuclease, Edc3 (enhancer of decapping). Xrn1 nuclease subsequently degrades the RNA. (**c**) Endonucleases can cut the mRNA, which leaves 3′ and 5′ end of each mRNA piece exposed to 3′ to 5′ (exosome) or 5′ to 3′ (Xrn 1) directions. (Figure from [36])

system using a dead or weakened viruses against measles 3000 years back [37]. Development of smallpox (1798) and cholera vaccines (1881) are roughly the beginning of modern vaccines. Vaccines started out by using an attenuated or dead virus to develop the immune response prior to an infection. The immune system would then recognize the pathogen upon exposure.

Instead of using the entire virus, surface proteins on the virus were found to be sufficient to generate an immune response. This changed the course of vaccine development [37, 38]. The antibodies generated against a particular foreign protein allows the body to recognize the virus and destroy it. Delivering the viral proteins to cells requires a delivery system.

As viruses are skilled at infecting the host, they are ideally suited to code for a novel protein derived from another virus. Adenoviruses isolated from humans cause mild cold-like symptoms. They have become an ideal delivery virus into human cells. The DNA of adenovirus is modified to carry any protein-coding sequence into the host cell. Adenoviruses are made defective in replication by removing their key replication proteins. Once inside the host cell, the DNA carried by the adenovirus is converted to mRNA and the protein of interest gets made inside the human cells. This new protein is foreign to the body and generates an immune response. In the current coronavirus pandemic, Johnson and Johnson, AstraZenca and Sputnik V all have an adenovirus-based SARS-CoV-2 vaccines. The spike protein (or portions of it) produced in the cells elicits antibody production by the immune system [39].

Many new vaccine technologies were developed when the world had to deal with the HIV-1 that causes AIDS starting in the 1980s. Different types of vaccines are designed around the idea of exposing the body to all or portion of viral proteins that are found on the virus surface. Current technologies include directly attaching proteins to nanoparticles or delivering the mRNA to cells, without the need for adenovirus for delivery.

The mRNA-based vaccines The idea of delivering mRNA corresponding to a protein of interest, has existed for many decades. Safe and effective methods for delivering RNA were being tested for decades. When lipid-based RNA delivery system based on lipid nanoparticles, LNP, were discovered, the potential of mRNA as vaccines could be realized [40].

The cationic lipid particles (fatty acids) coat the anionic RNA to protect it from nucleases and fuse with the cell membrane for uptake. A treatment that uses LNP to deliver small non-coding RNA (not mRNA) was developed for amyloid plaques. A drug, patisiran, that utilizes a LNP system for delivering small RNA into the cells was approved by FDA in 2018 after years of routine testing and trials.

The potential to develop mRNA-based medicines against infectious agents, cancers and allergies has been explored for decades [41]. RNA-based therapies have been expanding rapidly since the discovery of catalytic RNA in 1980s (Chap. 3, catalytic RNA). In the last decade, research on small non-coding RNA (Chap. 8, ncRNA) and CRISPR-Cas based genetic manipulation (Chap. 9, CRISPR) has expanded the potential for RNA-based therapies. Many new RNA-based therapies are currently being tested and the lipid-based delivery methods have been thoroughly researched.

Benefits of using mRNA for vaccines The mRNA is easy to synthesize, it has a short lifetime, and it does not become part of the host's genome. A small amount of mRNA is needed to make proteins to train the immune system. In a short time after delivery, the mRNA is degraded in the cell using the normal cellular processes (as discussed in Sect. 6.8). No additional viruses are introduced into the system when using mRNA. As the virus mutates its surface protein evade detection by the immune system. The mRNA nucleotide sequence corresponding to these mutations can be easily altered to create new booster vaccines.

History of mRNA vaccines In early 2010, mRNA technology was used to develop vaccines against influenza and rabies [42, 43]. In 2017, a biotech company Moderna reported an mRNA-based vaccine for the Zika virus in mice; a portion of membrane protein (M) and envelope (E) protein code were introduced into mice [44].

In the Zika virus, the E protein exists as 90 antiparallel homodimers on the surface. The E protein consists of three ectodomains (DI, DII, DIII) that are targeted by neutralizing antibodies. The mRNA for M-E protein was modified by using 1-methyl pseudouridine in place of uracil and contained $5'$ cap and $3'$ poly(A) tail. Modifications in uracil (by using pseudo uridine or 5-methyl cytidine) increases RNA's stability.

The designed mRNA was packaged into LNP for delivery to the cells. The lipid particles are made of an ionizable lipid, cholesterol, and polyethylene glycol. A solution of mRNA, in 1:3 ethanol and water, is mixed with lipids to generate lipid coated mRNA particles that are 80–100 nm in size. When the particles containing mRNA are injected, mRNA is brought into the cells via endosomes and released into the cytoplasm. mRNA are translated to their corresponding proteins which elicit the antibody response in mice. Normally, any injected RNA would immediately be degraded, or it may generate an immune response. The lipid particles protect the RNA from degradation and bring it inside the cell where the pH of the endosome causes the lipid particles to release the mRNA. The injection of mRNA covered in lipid produced less of an immune response than injection of the plain saline solution.

The success of any vaccines lies in generating the right type of immune response. The response has to be based on antibodies that are specific to a given virus. By 2017, mRNA-based immune cell activation was already being studied in rhesus macaques; the non-human primate studies are done before human trials. The promising results in these earlier trials allowed quick turnaround of mRNA-based vaccine development when the coronavirus pandemic started.

Why development of coronavirus vaccine seemed fast? By the time 2019 coronavirus pandemic started, the technology for delivery of mRNA vaccine and its ability to generate an immune response was well established. It had been extensively studied in mice and monkeys—aspects of vaccine trials that normally take years.

In case of the coronavirus, the spike protein that is present on the surface of the cell binds to the ACE-2 receptors of the host cells to gain entry. Many different coronaviruses had been studied already and target proteins to use for vaccines were already established. Therefore, mRNA designed to produce the spike protein (without the virus) could be tried quickly. The coronavirus was identified as the source of the disease in December 2019. By January 2020, coronavirus that causes Covid-19, SARS CoV-2, was sequenced and its sequence and organization of the genome were familiar and well understood. Coronaviruses are a single-stranded RNA viruses that have been studied for decades. Much was known about their genome architecture. SARS (severe acute respiratory syndrome, SARS-CoV-1) and MERS (Middle East respiratory syndrome) pandemics were both caused by coronaviruses in 2002 and 2012, respectively. Research on these earlier coronaviruses had started the vaccine development against these viruses. When the SAR-CoV-2 pandemic started researchers could analyze the sequence and structure of the spike protein and compare it to other coronavirus spike proteins [39]. Vaccine against coronaviruses (albeit different coronaviruses) were already being developed that allowed companies to put in the right messenger RNA against the SAR-CoV-2-specific spike proteins for vaccines trials.

The first doses of mRNA-based vaccines were ready for trials in March 2020. Decades of work on other coronaviruses, RNA and its delivery had taken place, including the prerequisite large animal studies. Even when the speed of vaccine development seems remarkable, the groundwork for these trials is laid for decades. Infections from other coronaviruses were expected given that it is a large family of viruses that had caused SARS and MERS already.

A quick and timely infusion of money from NIH and an established network for HIV vaccine research fueled the early vaccine trials. In addition, nearly 45,000 people volunteered to participate in the trials in a very short amount of time due to pandemic-based restrictions in place.

The speed of the coronavirus vaccine development was possible due to a lot of prior scientific research into coronaviruses (and other viruses), government funding, established networks, and support, and people ready for a "cure."

Future of mRNA vaccines Many more mRNA-based vaccines are on the horizon [45]. The technology developed for coronavirus vaccine is readily available to tailor to other viruses. The promise of mRNA-based treatments is immense as synthesis of RNA is quick and relatively straightforward. So far, two mRNA vaccines against SARS-CoV-2 (Pfizer and Moderna) have done well [39]. If a large population of people in the world can be vaccinated, then it is likely that the virus will not find new hosts to mutate, potentially bringing an end to the pandemic.

Take Home Message
- mRNA sequence and structures regulate rates of protein synthesis and contribute to protein folding.
- Our understanding of viral RNA has helped us to better understand the eukaryotic cells.
- The life cycle of mRNA shows the many checks and balances in the cells that determine the fate of the mRNA.
- Understanding mRNA-based vaccine technology shows that mRNA vaccines were developed with proper scientific protocols and have shown efficacy so far.

References

1. ENCODE Project Consortium. An integrated encyclopedia of DNA elements in the human genome. Nature. 2012;489:57–74.
2. Nilsen TW, Maroney PA, Robertson HD, et al. Heterogeneous nuclear RNA promotes synthesis of (2′,5′) oligoadenylate and is cleaved by the (2′,5′) oligoadenylate-activated endoribonuclease. Mol Cell Biol. 1982;2:154–60.
3. Kozak M. Regulation of translation via mRNA structure in prokaryotes and eukaryotes. Gene. 2005;361:13–37.
4. Sharp PA. The centrality of RNA. Cell. 2009;136:577–80.
5. Gentile GM, Wiedner HJ, Hinkle ER, et al. Alternate endings. The Scientist, January/February 38; 2020.
6. Nostrand EL, Freese P, Pratt GA, et al. A large-scale binding and functional map of human RNA-binding proteins. Nature. 2020;583:711–9.
7. Mauger DM, Cabral BJ, Presnyak V, et al. mRNA structure regulates protein expression through changes in functional half-life. Proc Natl Acad Sci. 2019;116:24075–83.
8. Mortimer SA, Kidwell MA, Doudna JA. Insights into RNA structure and function from genome-wide studies. Nat Rev Genet. 2014;15(7):469–79.
9. Bevilacqua P, Ritchey L, Zhao S, Assmann S. Genome-wide analysis of RNA secondary structure. Annu Rev Genet. 2016;50:235–66.
10. Wan Y, et al. Landscape and variation of RNA secondary structure across the human transcriptome. Nature. 2014;505:706–9.
11. Cruz JA, Westhof E. The dynamic landscapes of RNA architecture. Cell. 2009;136:604–9.
12. Martin KC, Ephrussi A. mRNA localization: gene expression in the spatial dimension. Cell. 2009;136:719–30.
13. Potter KD, Merlino NM, Jacobs T, Gollnick P. TRAP binding to the Bacillus subtilis trp leader region RNA causes efficient transcription termination at a weak intrinsic terminator. Nucleic Acids Res. 2011;39:2092–102.
14. Link K, Breaker R. Engineering ligand-responsive gene-control elements: lessons learned from natural riboswitches. Gene Ther. 2009;16:1189–201.

15. Kushner S. mRNA decay in prokaryotes and eukaryotes: different approaches to a similar problem. Life. 2004;56:585–94.
16. Hocine S, Singer RH, Grünwald D. RNA processing and export. Cold Spring Harb Perspect Biol. 2010;12:a000752.
17. Rotival M, Quach H, Quintana-Murci L. Defining the genetic and evolutionary architecture of alternative splicing in response to infection. Nat Commun. 2019;10:1671.
18. Herzel L, Ottoz DSM, Alpert T, Neugebauer KM. Splicing and transcription touch base: co-transcriptional spliceosome assembly and function. Nat Rev Mol Cell Biol. 2017;18:637–50.
19. Cole C. Choreographing mRNA biogenesis. Nat Genet. 2001;29:6–7.
20. Fehr AR, Perlman S. Coronaviruses: an overview of their replication and pathogenesis. In: Maier H, Bickerton E, Britton P, editors. Coronaviruses. Methods in molecular biology, vol. 1282. New York, NY: Humana Press; 2015.
21. Jin Y, Yang H, Ji W, et al. Virology, epidemiology, pathogenesis, and control of COVID-19. Viruses. 2020;12(4):372.
22. Luciw PA. Human immunodeficiency viruses and their replication. In: Fields BN, editor. Virology. 3rd ed. Philadelphia: Lippincott-Raven; 1996.
23. Simmonds P. Pervasive RNA secondary structure in the genomes of SARS-CoV-2 and other coronaviruses. Ecol Evol Sci. 2020;11:e01661–20.
24. Watts JM, Dang KK, Gorelick RJ, et al. Architecture and secondary structure of an entire HIV-1 RNA genome. Nature. 2009;460:711–6.
25. Samuel CE. RNA editing minireview series. J Biol Chem. 2003;278:1389–90.
26. Brennicke A, Marchfelder A, Binder S. RNA editing. FEMS Microbiol Rev. 1999;23:297–316.
27. Butler JS. The Yin and Yang of the exosome. Trends Cell Biol. 2002;12:90–6.
28. Fasken M, Corbett A. Process or perish: quality control in mRNA biogenesis. Nat Struct Mol Biol. 2005;12:482–8.
29. Green DM, Johnson CP, Hagan H, et al. The C-terminal domain of myosin-like protein 1 (Mlp1p) is a docking site for heterogeneous nuclear ribonucleoproteins that are required for mRNA export. Proc Natl Acad Sci.U S A. 2003;100:1010–5.
30. Vinciguerra P, Iglesias N, Camblong J, Zenklusen D, et al. Perinuclear Mlp proteins downregulate gene expression in response to a defect in mRNA export. EMBO J. 2005;24: 813–23.
31. Balagopal V, Parker R. Polysomes, P bodies and stress granules: states and fates of eukaryotic mRNAs. Curr Opin Cell Biol. 2009;21:403–8.
32. Carmody SR, Wente SR. mRNA nuclear export at a glance. J Cell Sci. 2009;122:1933–7.
33. Hilleren P, Parker R. Mechanisms of mRNA surveillance in eukaryotes. Annu Rev Genet. 1999;33:229–60.
34. Parchure A, Munson M, Budnik V. Getting mRNA-containing ribonucleoprotein granules out of a nuclear back door. Neuron. 2017;96:604–15.
35. Costa DI, Buchanan CN, Zdradzinski MD, et al. The functional organization of axonal mRNA transport and translation. Nat Rev Neurosci. 2021;22:77–91.
36. Garneau N, Wilusz J, Wilusz C. The highways and byways of mRNA decay. Nat Rev Mol Cell Biol. 2007;8:113–26.
37. Glynn I, Glynn J. The life and death of smallpox. Cambridge, UK: Cambridge University Press; 2004.
38. Greenwood B. The contribution of vaccination to global health: past, present and future. Philos Trans R Soc Lond Ser B Biol Sci. 2014;369:20130433.
39. Krammer F. SARS-CoV-2 vaccines in development. Nature. 2020;586:516–27.
40. Cross R. Without these lipid shells, there would be no mRNA vaccines for COVID-19. Chem Eng News. 2021;99:8.

41. Liang F, Lindgren G, Lin A, et al. Efficient targeting and activation of antigen-presenting cells in vivo after modified mRNA vaccine administration in Rhesus Macaques. Mol Ther. 2017;25: 2635–47.
42. Hekele A, Bertholet S, Archer J, et al. Rapidly produced SAM(a) vaccine against H7N9 influenza is immunogenic in mice. Emerg Microbes Infect. 2013;2:e52.
43. Petsch B, Schnee M, Vogel AB, et al. Protective efficacy of in vitro synthesized, specific mRNA vaccines against influenza A virus infection. Nat Biotechnol. 2012;30:1210–6.
44. Richner JM, Himansu S, Dowd KA, et al. Modified mRNA vaccines protect against zika virus infection. Cell. 2017;168:1114–25.
45. Pardi N, Hogan M, Porter F, et al. mRNA vaccines—a new era in vaccinology. Nat Rev Drug Discov. 2018;17:261–79.

Riboswitches: Sensors and Regulators

7

Sriya Sharma and Neena Grover

Contents

Keywords

Riboswitches · RNA sensors · RNA regulation · RNA thermodynamic control · RNA kinetic control

What You Will Learn

Riboswitches are structures in mRNA that change shape upon binding a metabolite. A specific set of interactions occur between the RNA and its cognate ligand to

(continued)

S. Sharma · N. Grover (✉)
Department of Chemistry and Biochemistry, Colorado College, Colorado Springs, CO, USA
e-mail: s_sharma@coloradocollege.edu; NGrover@ColoradoCollege.edu

© Springer Nature Switzerland AG 2022
N. Grover (ed.), *Fundamentals of RNA Structure and Function*, Learning Materials in Biosciences, https://doi.org/10.1007/978-3-030-90214-8_7

differentiate it from other similar molecules in the cell. The binding of a metabolite influences RNA structures and hence, the downstream events regulating gene expression. Riboswitches have an aptamer binding domain that senses the ligand concentration and an expression platform that influences the fate of the mRNA. The mechanisms of gene regulation used by riboswitches will be discussed in this chapter. Riboswitches as means of gene regulation are often seen in bacteria, some of which are harmful to human health and hence, targeting these structures may be a path to new antibiotics.

Learning Objectives
After reading this chapter you should be able to:

- Explain the role of riboswitches in gene regulation.
- Describe the functions of the aptamer domains and expression platforms.
- Identify some modes of ligand binding and associated structural changes that lead to different mRNA outcomes.

7.1   Introduction

Organisms control gene expression in response to their environments. Gene expression was thought to be exclusively controlled by proteins due to their large, complex, and variable structures. The discovery of catalytic function in group I introns, with their requirement of an exogenous guanosine and magnesium ions, set the stage for RNA's ability to bind small molecules [1]. Artificial evolution experiments (in vitro evolution or SELEX) subsequently generated small RNA (aptamers) that could selectively bind to small molecules [2]. The creativity of a few scientists to look for aptamer-like sequences in the known RNA sequences led them to discover conserved regions, and associated structures, in the untranslated regions of mRNA. These RNA structures were then shown to bind metabolites to self-regulate downstream events, often altering fate of the mRNA downstream of the binding event [3–7].

The discovery of riboswitches in bacterial mRNA, and subsequently in most organisms, was possible due to improvements in computational approaches and a revolution in genome sequencing.

Riboswitches are structured regions often found in the noncoding region of mRNA that serves as sensors of small molecules. Each riboswitch binds to a particular metabolite in a small range of concentrations to regulate genes related to its own metabolism, use, or transport.

The most abundant riboswitches recognize coenzymes and related compounds: thiamine pyrophosphate (TPP), S-adenosyl methionine (SAM), vitamin B_{12}, flavin mononucleotide (FMN), among others. The second largest group recognizes purines (adenine, guanine) and molecules related to it (e.g., prequeuosine, cyclic-di-AMP). A sizable number bind to amino acids (e.g., glycine, lysine, or glutamate). Some ion sensors are also labeled riboswitches (e.g., F^- and Mg^{2+}). Note that many of the molecules that bind to riboswitches have some nucleotide or nucleotide-like components, others are amino acid side chains that often interact with RNA (RNA-protein interactions), perhaps hinting at an ancient mechanism of self-regulation present in primarily RNA-based world.

Riboswitches contain highly conserved sequences that are found across all three domains of life, with most found in bacteria. Riboswitches form directly upstream of the gene they regulate. In bacteria, riboswitches are found in the 5′ UTR region of mRNA. The riboswitch binds very specifically to a cognate metabolite of the regulated gene via numerous hydrogen bonds, stacking and packing interactions. For example, a SAM-binding riboswitch must discriminate it from its metabolic byproduct, S-adenosylhomocysteine (SAH) which is toxic to the cell but very similar structurally (Fig. 7.1). A positive charge on sulfur and its interaction with the carbonyl carbons of two uracil nucleotides (not phosphates) and the methyl group on sulfur allows a 100-fold discrimination between the two molecules by SAM-I riboswitch. The specific contacts between the metabolite and the RNA allow for stabilizing neighboring interactions in RNA structures. The structures that perform the recognition of a metabolite vary among organisms—for example, there are several different classes of riboswitches that bind SAM or SAH, showing the versatility of different RNA structures to accomplish the same goal. The ligand binding energy is used to change the shape of the RNA.

Most commonly, riboswitches are activated through a negative feedback loop in which elevated levels of a metabolite leads to repression of the gene(s) responsible for its synthesis. A salient feature of riboswitches is their ability to switch between (at least) two conformational states in response to their metabolite's concentration.

Riboswitches are composed of two domains: an aptamer binding domain and an expression platform (Fig. 7.2) [3–7]. Upon binding to the appropriate metabolite, a riboswitch undergoes a conformational change in the expression platform which is responsible for the subsequent gene regulation. Many riboswitches have a sequence that can either base pair within the aptamer domain or form an alternate structure with the expression platform. The predominant conformational states of the riboswitch depend on the concentration of their cognate ligand. In some riboswitches, the expression platform is non-discrete from the aptamer domain. Although many classes of riboswitches have been discovered, many other classes are likely to exist. Candidate riboswitch sequences are being identified in genomic databases using sequence and secondary structure features seen in the 5′-UTR of mRNA. The ligands that bind these have yet to be discovered.

Among bacteria, riboswitches are predominately used by gram-positive bacteria as a mechanism for regulating gene expression. In some bacteria, riboswitches are responsible for controlling anywhere from 2 to 4% of gene regulation. The existence of bacteria-

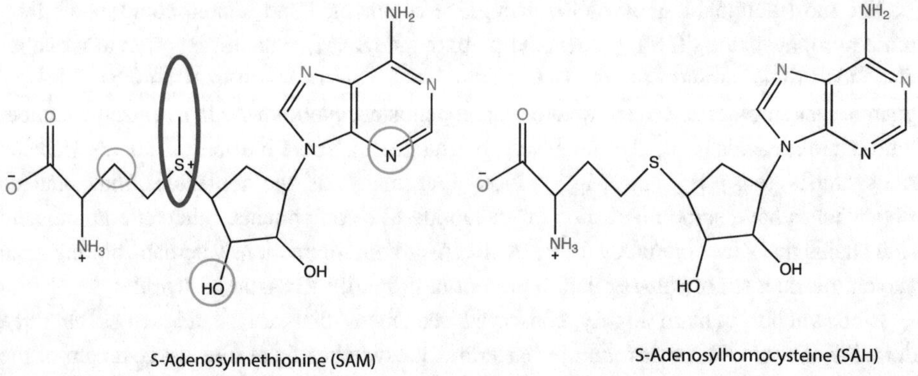

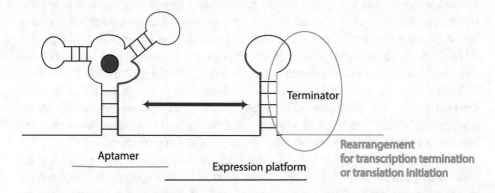

S-Adenosylmethionine (SAM)   S-Adenosylhomocysteine (SAH)

Fig. 7.1 Metabolites make specific contacts with RNA. A metabolite must be distinguished from other similar molecules in a cell. The RNA binds the metabolites through many hydrogen-bond donor (blue atoms) and acceptor sites (red atoms). In addition, stacking interactions between rings and other packing interactions (light blue circles) within RNA allow for specific recognition of a particular ligand. The various hydrogen-bond donor and acceptors for *S*-adenosylmethionine (SAM) and some of its van der Waal interaction sites are colored or circled. The positive charge on sulfur (dark blue ellipse) interacts with two different O2 sites on binding site uracils to discriminate it from its toxic byproduct *S*-adenosylhomocysteine (SAH) [3]

Fig. 7.2 A schematic of the aptamer domain and the expression platform. The schematic above depicts the binding of an aptamer (red sphere) alters the availability of light blue region (purple with light blue or purple with purple). In general, riboswitches work by altering RNA structures near the aptamer domain as a means to regulate genes

specific RNA regulatory structures opens up the possibility of designing antibiotics that target riboswitches.

In eukaryotes, the thiamine pyrophosphate riboswitch (TPP) is responsible for changes in gene expression through alternative splicing [8]. In some fungi, TPP causes a structural change in the mRNA, either creating access to or occluding a splice site. The presence or

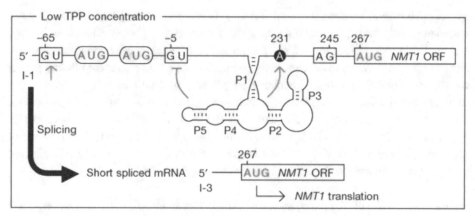

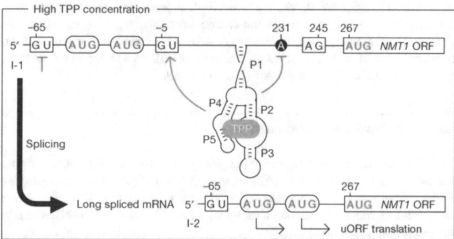

Fig. 7.3 The binding of thiamine pyrophosphate controls splicing. The *NMT-1* gene expression is repressed by thiamine pyrophosphate (TPP) in *N. crassa*. The thiamine pyrophosphate-binding riboswitch resides in the intron near the 5′ terminus. Thiamine binding (K_d ~300 pM) causes alternate splicing of the *NMR-1* gene. The key splicing determinants (GU at position −65 or −5 positions; and branch point adenine, at position 231) are differently available under high or low concentrations of TPP. (Figure from [8])

absence of TPP is thus responsible for determining which mRNA isoforms are produced (Fig. 7.3).

7.2   The Aptamer Domain and the Expression Platform

The aptamer domain The aptamer domain is a region of RNA that has a highly conserved sequence and binds to a particular metabolite with high specificity, while rejecting others that may be structurally similar (Fig. 7.1). This domain makes precise hydrogen bonding, electrostatic, and stacking interactions, along with utilizing shape complementarity with its metabolite, to allow for selectivity [3–15]. Mutations in even one nucleotide can change the ability of a riboswitch to properly bind its intended ligand. For example, in purine riboswitches, only one nucleotide differentiates the binding pocket of adenine riboswitches from guanine riboswitches; mutation of this nucleotide can diminish the specificity essential for proper gene regulation in response to nucleotide levels. In some riboswitches, the entire molecule is recognized as seen in purine riboswitches (Fig. 7.4) [9, 11]. In others, the binding occurs at the peripheral portions of the molecule while measuring the total length of the molecule, as seen in the thiamine pyrophosphate riboswitch (Fig. 7.5) [12].

Expression Platform An expression platform is the region of RNA that undergoes a conformational change (allostery) upon ligand binding, leading to downstream effects on gene expression. The expression platform of a riboswitch shows a great deal of diversity across species. The change in conformations of RNA may or may not be reversible due to the short lifetime of mRNA in the cell. When irreversible, it leads to degradation of RNA. In this case, a ligand is creating a fuse instead of a switch.

7.3   Mechanisms of Regulation

The two primary mechanisms for controlling gene expression are cis-transcription attenuation and cis-translation repression, implying self-regulation of the mRNA. Only one class of riboswitches uses cis self-cleavage to regulate its gene expression. Some riboswitches use a trans transcription termination mechanism, meaning regulation of an alternate gene (this mechanism is not discussed here).

Cis-transcription attenuation   If regulation and transcription were to occur simultaneously then the folding of the expression platform of RNA needs to be faster than its elongation to affect further gene expression. The two primary RNA conformations that regulate transcription in the riboswitches are the formation of an anti-terminator stem or a terminator stem.

In some riboswitches, in low concentrations of ligand, a normal expression of a gene is seen. In high concentrations of a ligand, the ligand binds to the RNA to cause repression of gene expression. In low level of ligands, an anti-terminator stem forms. In the presence of high ligand concentrations, the terminator helix forms downstream of the aptamer domain, leading to the disassociation of the RNA polymerase, thus stopping its transcription.

Different riboswitches can turn transcription on or off depending on the sensor molecule binds to it and the genes that are being regulated. The genetic context of the riboswitch dictates whether the ligand binding stabilizes or destabilizes the formation of the terminator helix.

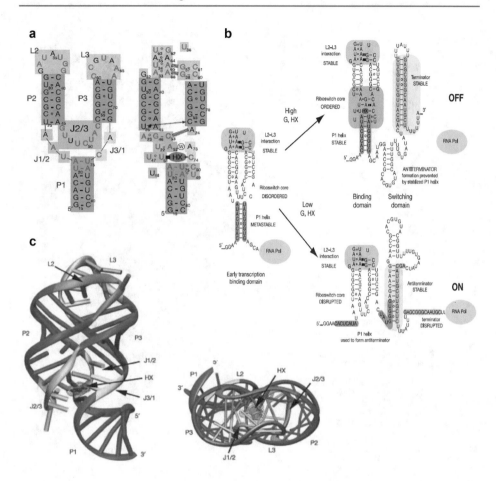

Fig. 7.4 Guanine-responsive riboswitch. (**a**) Secondary structures of guanine riboswitch with and without guanine bound; guanine is indicated as red sphere (HX, hypoxanthine—a purine precursor molecule); nucleotide shown in red are often conserved. (**b**) In the presence of high concentration of guanine, P1 helix (green) is stabilized to form an terminator hairpin which stops further transcription of the RNA (top); in low concentration of guanine, the P1 helix is destabilized, leading to an anti-terminator hairpin to form, allowing further transcription of the mRNA. (**c**) The crystal structure of guanine riboswitch with bound guanine. (**d**) A top down view of the structure in (**c**). (Figure from [9])

Cis-translation attenuation Riboswitches that control translation affect the Shine-Dalgarno ribosome binding site (RBS) or access to the start codon (AUG) for translation. This mechanism of control relies on the use of sequestering sequences that either remain within the aptamer region or form a hairpin with the Shine-Dalgarno sequence making in inaccessible, thus, preventing ribosome binding.

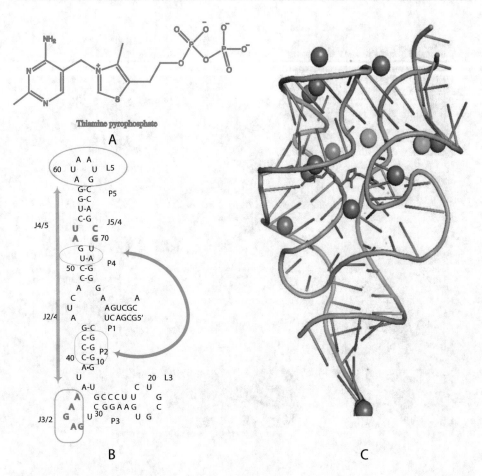

Fig. 7.5 Thiamine pyrophosphate riboswitch. The thiamine pyrophosphate (**a**) molecule binds in the junction regions of the RNA (**b**, **c**). TPP is shown as sticks in the crystal structure in (**c**). The pyrophosphate is bound to RNA (with metal ions neutralizing the charge on phosphates) in J4/5 and J5/4 (residues shown in red). The aminopyrimidine ring makes several interactions in J3/2 (residues in red). The ring stacks between two purines (G and A) and hydrogen bonds to another, allowing folding of the RNA. Mn^{2+} and Mg^{2+} ions are shown as purple and green spheres in (**c**). (Figure made using PDB file 2HOJ in PyMol [12])

Commonly, gram negative bacteria use the translation control mechanism whereas gram-positive bacteria primarily use the transcription attenuation mechanism for gene regulation. In different organisms, the same aptamer may regulate gene expression through a completely different mechanism.

7.4   Structural Organization within Riboswitches

The three-dimensional structures of many riboswitches have been solved [6–20]. A great deal of variability is seen in the riboswitch architecture, but some common themes have emerged.

All riboswitches are currently divided into two main categories: type I and type II [15–19]. Type I riboswitches have a tightly folded structure for metabolite binding that undergo small local conformational changes upon ligand binding. The active site is mostly pre-formed in the apo-RNA. For example, the purine riboswitches are characterized by a single binding pocket formed by a pre-established global fold. This limits ligand-induced changes to a very small region.

Type II riboswitches have a primarily unfolded structure that folds into the functional structure upon metabolite binding. A thiamine pyrophosphate riboswitch has a binding pocket that is split into two distinct sites, one that binds the thiamine part and other side that recognizes the pyrophosphate moiety. When thiamine pyrophosphate binds, the global architecture of RNA changes (Fig. 7.5).

Most riboswitch structures are comprised of pseudoknots, junctions, or both (Fig. 7.6). Helical structures are responsible for structural stabilization in all classes. Mixed riboswitches contain components of both the pseudoknotted and junctional subdivisions. The tertiary interactions often further from the binding site play a crucial role in stabilizing the overall folded structures, as seen in the guanine binding riboswitch loop L2 and L3 interactions (Fig. 7.4).

Fig. 7.6 Ligand binding sites. A pseudoknot or a junction region bind the ligand to cause changes in the expression domain structures [19]

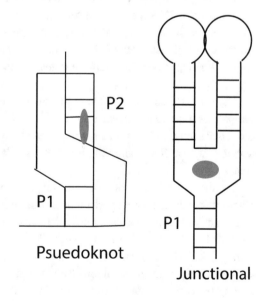

Psuedoknot

Junctional

7.5   Kinetic and Thermodynamic Control of Riboswitches

The conformation flexibility of RNA is at the heart of riboswitch functions. The RNA structures that form in riboswitches control the fate of mRNA by either the rate at which they form (kinetically) or by shifting the equilibrium between different structures (thermodynamics) upon ligand binding.

In kinetic control, the rate of structure formation determines the riboswitch response. Riboswitches must form a stable conformation within the expression platform at a rate faster than RNA polymerase elongation of the mRNA strand or the binding of the ribosome at the Shine-Dalgarno sequence. The kinetic model of RNA folding is illustrated for the fluoride riboswitch in Fig. 7.7 [21].

Riboswitches which regulate gene expression through transcription attenuation have greater kinetic constraints than those which regulate gene expression through translation inhibition. During transcriptional control, the expression platform of a riboswitch must fold faster than RNA elongation (50 nt/s) to affect further gene expression. While this process is assisted by intrinsic RNA polymerase pause sites within mRNA strands, it heavily relies on a ligand concentration available to be greater than the disassociation constant (K_d) of the riboswitch. The ligand bound conformation need to be stabilized long enough to induce a conformational change in the expression platform.

The riboswitch is primarily under thermodynamic control when the binding affinity and cellular concentrations of a ligand determine the equilibrium between conformations [3–7, 21, 22].

Riboswitch structures are affected by the presence of competing conformations. In the unbound state, RNA may take on several conformations and only some of these are able to bind the corresponding ligand. While some structures are thermodynamically favored over others, an equilibrium may exist between several unbound structures of similar stability. Ligand binding drives the equilibrium toward structures which are capable of binding the ligand [22].

The presence and concentration of ions as well as environmental conditions (such as, temperature) influence the thermodynamic stability of RNA conformations. Magnesium and potassium ions neutralize the negative charges on the backbone of RNA and bind to specific sites on RNA, allowing it to form stable structures that bind their corresponding ligand.

For the lysine riboswitch, a short-range of magnesium ion concentrations tunes the riboswitch activity between kinetic or thermodynamic control. The presence of magnesium ions changes the predominant riboswitch structure. A higher concentration of magnesium ions lead to the formation of a pre-folded riboswitch structure capable of binding lysine tightly (K_d ~1 μM) and controlling gene regulation thermodynamically. A lower concentrations of magnesium ions lead to an unfolded riboswitch structure which requires a greater concentration of lysine to binds to it (K_d ~180 μM), using an induced fit mechanism to control gene expression kinetically (Fig. 7.8).

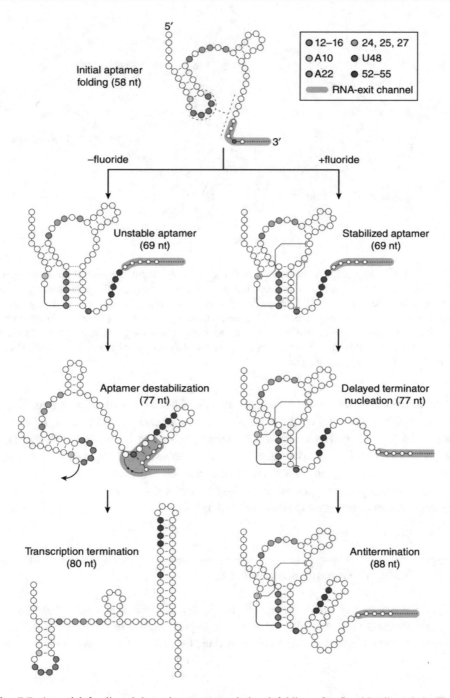

Fig. 7.7 A model for ligand-dependent co-transcriptional folding of a fluoride riboswitch. The folding of RNA begins (right side) when fluoride binds to the RNA. Upon fluoride binding, the aptamer is stabilized via specific interactions that cause a delay in early stages of folding of the intrinsic terminator hairpin. Thus, an anti-termination helix forms. In the absence of fluoride (left

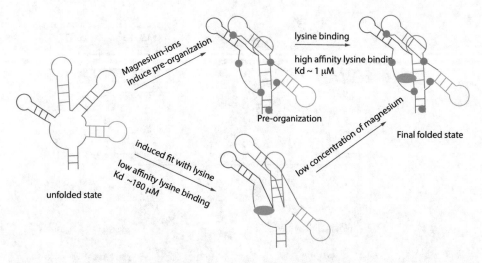

Fig. 7.8 The affinity of the riboswitch for its ligand is influenced by its environment. A schematic depiction of the lysine binding to its riboswitch under different ionic conditions is shown (the true folding intermediates are in [22]). The affinity of lysine for the riboswitch is tuned by a very narrow range of magnesium concentration. In high concentrations of magnesium (~2 mM), riboswitch can adopt a folded conformation ready to bind lysine with high affinity. At low concentrations of magnesium (~0.5 mM), the riboswitch binds to lysine with low affinity and the structures form with the assistance of the lysine [22]

In some riboswitches, a decrease in the amount of ligand needed to cause a conformational change is obtained by the utilizing two riboswitch regions upstream of a gene (discussed below for glycine riboswitch). Riboswitches can be stacked such that both riboswitches respond to the same ligand or they can sense different ligands. Riboswitches may also contain two aptamers with the same expression platform.

Translation regulation does not have tight temporal constrains as the entire mRNA molecule is transcribed prior to ribosome binding.

7.6   Tandem Glycine Riboswitches

Glycine concentrations are maintained tightly in the cells; elevated levels of glycine are harmful for the formation of the cell walls and impact survival of the cells [23–26]. The glycine riboswitch contains tandem aptamer domains that bind to glycine to turn on the expression of glycine degrading genes within the *gcvT* operon.

Fig. 7.7 (continued) side), the terminator hairpin forms and disrupts the pseudoknot structure to trigger transcription termination [21]

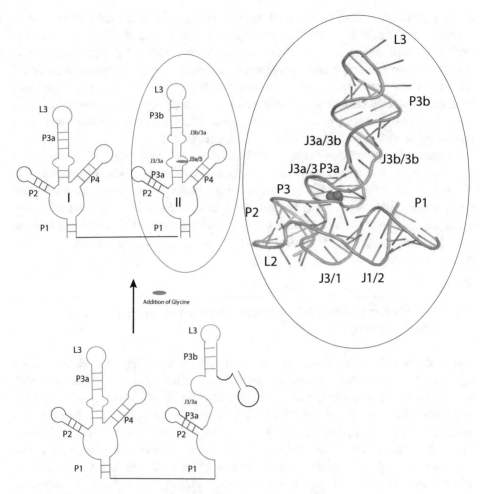

Fig. 7.9 The glycine riboswitch. Tandem aptamers, marked I and II, exist for glycine riboswitch. In the absence of glycine, the aptamer forms a terminator stem (red). In the presence of glycine, an anti-terminator stem forms [25]. The crystal structure (2.95 Å resolution) of a glycine bound domain II bound is shown. (The figure was made using PDB file 3OWZ in PyMol)

The structure of the tandem aptamers of the glycine riboswitch is highly conserved. In the presence of glycine, the conformation of the RNA changes to form an anti-terminator stem (Fig. 7.9). In the absence of glycine, a terminator stem forms in the expression platform, controlling gene expression through transcription attenuation. The two aptamer regions are separated by a conserved linker. An additional linker is implicated in the dimerization of the two aptamers and lies upstream of aptamer 1.

Glycine riboswitches are junctional, made up of three helical stems connected to a junction region. Glycine binds adjacent to the three-way junction which pushes an adenine from the binding pocket into the junction to stabilize the P1 helix. Glycine binds to the aptamer region by fitting into a tight binding pocket which can only accommodate structures of the same size; glycine forms hydrogen bonds with residues in the binding pocket. Magnesium ions stabilize glycine binding by neutralizing the negative charge on the alpha-carboxylate ion. Three predominant conformational states have been observed in this riboswitch: two unbound conformations and one bound. Without magnesium ion, glycine riboswitches form a conformation unable to bind glycine. Magnesium ions tighten folding by stabilizing charges on the RNA backbone allowing a shift in equilibrium toward conformations that are capable of binding glycine (Fig. 7.9).

The tandem aptamer allows cooperative binding because the binding of glycine to one aptamer was seen to significantly increase ligand binding at the second aptamer. The binding was described with a Hill coefficient of 1.64. Binding to both aptamers is necessary for optimizing riboswitch activity because disruptions to binding in either aptamer, led to decreased gene expression. Although noncooperative models for glycine binding have been also been proposed, the current consensus leans toward a cooperative model of glycine binding.

7.7 GlmS Riboswitches Regulate Gene Expression Through Self-Cleavage of mRNA

The glmS riboswitch class is the only known riboswitch to undergo self- cleavage upon ligand binding, classifying it as both a riboswitch and a ribozyme [27–33]. The riboswitch lies upstream of the glutamine-fructose-6-phosphate amidotransferase gene and is activated by glucosamine-6-phosphate (GlcN6P). This gene encodes for proteins implicated in cell wall biosynthesis and is, thus, essential to the growth and survival of bacteria. High levels of GlcN6P lead to deactivation of the glmS gene through cleavage at a single nucleotide site downstream of the riboswitch. GlcN6P is directly involved in the catalytic mechanism of the glmS ribozyme and increases the rate of reaction by about 10^5. Unlike most riboswitches, GlcN6P does not cause significant conformational changes in the expression platform but acts as a cofactor essential to ribozyme function.

The glmS riboswitch has a highly structured and pre-formed binding pocket and active site that does not undergo significant conformational changes upon ligand binding. The structure is considered mixed as the tertiary structure is characterized by helical stacks as well as pseudoknots (Fig. 7.10) [28]. The ribozyme is made up of four main helical domains (P1–P4) whose tertiary interactions stabilize the active site and binding pocket. The P2 domain of the riboswitch is essential for catalytic activity and the P2.1/P2 domain are responsible for metabolite recognition. These two domains pack tightly into a double pseudoknot, stabilized by base triples, which form both the active site and ligand binding pocket. The P3/P4 domains, while not essential for catalysis, improve the rate of catalysis and are expected to stabilize the ribozyme's conformation through tertiary interactions.

The active site is aligned for cleavage reaction (pre-cleavage state). The active site nucleotides are stabilized by stacking interactions as well as non-canonical base

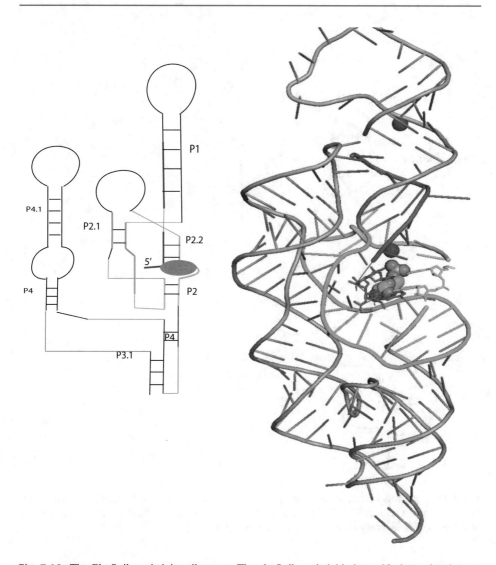

Fig. 7.10 The GlmS riboswitch is a ribozyme. The glmS riboswitch binds to a N-glycosylated sugar (green/red spheres) that participates in the cleavage mechanism. (**a**) A cartoon depiction of the secondary structure corresponding to the tertiary structure. (**b**) The tertiary structure (3 Å resolution) shows the double psuedoknotted area (green/teal) that forms in the middle of the riboswitch by P2 and P2.1. The substrate strand (purple) cleavage site is on the 5′ end and is buried in the middle (purple sticks) near the coenzyme. Magnesium ions (blue spheres) are not positioned for cleavage in the active site, but one is in the vicinity [28]. (Figure made using PDB file 3G8T in PyMol)

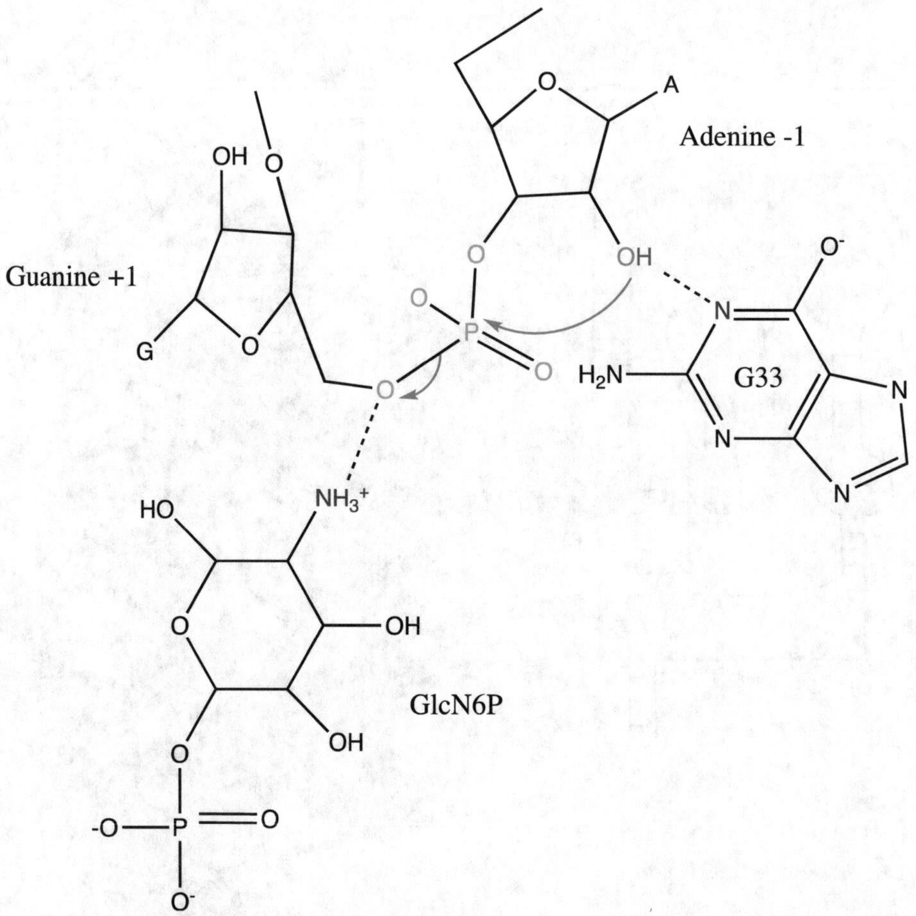

Fig. 7.11 A potential catalytic mechanism of GlmS riboswitch involves the GlcN6P as a cofactor. GlmS riboswitch uses RNA cleavage as a mechanism of mRNA regulation. GlcN6P binds in the active site and is involved in stabilizing the transition state [31]

interactions. GlcN6P binds to the ligand binding site through phosphate–magnesium interactions along with hydrogen bonding between active site nucleotides and water molecules. GlcN6P also directly stacks with the active site guanine and an adenosine nucleotide, supporting the active site conformation. The binding pocket buries most of the GlcN6P ligand; however, the phosphate end remains open to solvent, stabilized by hydrated magnesium ions. These Mg^{2+} ions are essential for optimizing activity through structural stabilization but do not directly impact the catalytic mechanism.

The catalytic mechanism is shown in Fig. 7.11 and further discussed in the Chap. 3 on small catalytic RNA. The resulting mRNA strand is unstable and is targeted by a RNase for degradation. The GlcN6P is an important cofactor in this reaction.

The glmS riboswitch is regulated by multiple metabolites involved in hexose metabolism. These metabolites have been found to both activate and inhibit ribozyme activity in order to regulate gene expression suggesting that the glmS riboswitch responds to complex downstream metabolic pathways related to its primary metabolite.

7.8 Medical Implications

Riboswitches modulate expression of genes vital to the growth and survival of bacteria. Understanding the mechanisms by which riboswitches exert control may have many medical implications. Targeting of the pathogenic bacterial riboswitches with antibiotics to disrupt key bacterial processes may prove effective in disease mitigation [34–36].

Riboswitches are ideal candidates for antibacterial drug development because of their high binding specificities and their lack of distribution in the human genome [7, 18, 34]. By understanding the mechanisms of binding, specific molecules can be developed which primarily bind the intended riboswitch. These compounds can theoretically be used as antibiotics because they target specific RNA conformations supposedly not found in the human genome. This process is nontrivial as the metabolites used by riboswitches are also used by humans and any analogs of metabolites are likely to bind to and inhibit human cellular processes. In addition, targeting pathogenic species may unintentionally target those bacteria that are beneficial to human health.

Take Home Message
- In bacteria, 2–4% of gene regulation occurs via the riboswitch structural rearrangements. While in eukaryotes, thiamine pyrophosphate binding to alter splice-site selection is the primary example of riboswitches. Further analysis of genomes is identifying putative riboswitches whose ligands have yet to be identified.
- In a riboswitch, the binding of a particular metabolite causes changes in RNA structure that determine the fate of the mRNA. The plasticity of RNA structures is key to the riboswitch-based gene regulation.
- The diversity of RNA structures that can bind to small molecule with specificity shows the verstality of RNA structures. The detailed biochemical knowledge of interactions between small molecules and RNA is likely to provide clues for developing new form of antibiotics against pathogenic bacteria.

References

1. Bass BL, Cech TR. Specific interaction between the self-splicing RNA of tetrahymena and its guanosine substrate: implications for biological catalysis by RNA. Nature. 1984;308:820–6.

2. Osborne SE, Ellington AD. Nucleic acid selection and the challenge of combinatorial chemistry. Chem Rev. 1997;97:349–70.
3. Breaker RR. Complex riboswitches. Science. 2008;319:1795–7.
4. Winkler WC, Breaker RR. Regulation of bacterial gene expression by riboswitches. Annu Rev Microbiol. 2005;59:487–517.
5. Winkler WC, Nahvi A, Roth A, Collins JA, Breaker RR. Control of gene expression by a natural metabolite-responsive ribozyme. Nature. 2004;428:281–6.
6. Roth A, Breaker RR. The structural and functional diversity of metabolite-binding riboswitches. Annu Rev Biochem. 2009;78:305–34.
7. Breaker RR. Riboswitches and the RNA world. Cold Spring Harb Perspect Biol. 2012;4:63–77.
8. Cheah MT, Wachter A, Sudarsan N, Breaker RR. Control of alternative RNA splicing and gene expression by eukaryotic riboswitches. Nature. 2007;447:497–500.
9. Batey R, Gilbert S, Montange R. Structure of a natural guanine-responsive riboswitch complexed with the metabolite hypoxanthine. Nature. 2004;432:411–5.
10. Mandal M, Breaker RR. Adenine riboswitches and gene activation by disruption of a transcription terminator. Nat Struct Mol Biol. 2004;11(1):29–35.
11. Liberman JA, Wedekind JE. Base ionization and ligand binding: how small ribozymes and riboswitches gain a foothold in a protein world. Curr Opin Struct Biol. 2011;21(3):327–34.
12. Edwards TE, Ferré-D'Amaré AR. Crystal structures of thi-box riboswitch bound to thiamine analogs reveal adaptive RNA-small molecule recognition. Structure. 2006;2006:1459–68.
13. Batey RT. Structures of regulatory elements in mRNAs. Curr Opin Struct Biol. 2006;16:299–306.
14. Mandal M, Boese B, Barrick JE, Winkler WC, Breaker RR. Riboswitches control fundamental biochemical pathways in bacillus subtilis and other bacteria. Cell. 2003;113:577–86.
15. Mandal M, Lee M, Barrick JE, Weinberg Z, Emilsson GM, Ruzzo WL, Breaker RR. A glycine-dependent riboswitch that uses cooperative binding to control gene expression. Science. 2004;306:275–9.
16. Montage RK, Batey RT. Riboswitches: emerging themes in RNA structure and function. Annu Rev Biophys. 2008;37:117–33.
17. Garst AD, Batey RT. A switch in time: detailing the life of a riboswitch. Biochim Biophys Acta. 2009;1789:584–91.
18. Pavlova N, Kaloudas D, Penchovsky R. Riboswitch distribution, structure, and function in bacteria. Gene. 2019;708:38–48.
19. Peselis A, Serganov A. Themes and variations in riboswitch structure and function. Biochim Biophys Acta. 2014;1839:908–18.
20. Reiter NJ, Chan CW, Mondragón A. Emerging structural themes in large RNA molecules. Curr Opin Struct Biol. 2011;21:319–26.
21. Watters KE, Strobel EJ, Yu AM, et al. Cotranscriptional folding of a riboswitch at nucleotide resolution. Nat Struct Mol Biol. 2016;23:1124–33.
22. McCluskey K, Boudreault J, St-Pierre P, et al. Unprecedented tunability of riboswitch structure and regulatory function by sub-millimolar variations in physiological Mg2+. Nucleic Acids Res. 2019;47:6478–87.
23. Huang L, Serganov A, Patel DJ. Structural insights into ligand recognition by a sensing domain of the cooperative glycine riboswitch. Mol Cell. 2010;40:774–86.
24. Lipfert J, Sim AYL, Herschlag D, Doniach S. Dissecting electrostatic screening, specific ion binding, and ligand binding in an energetic model for glycine riboswitch folding. RNA. 2010;16:708–19.
25. Ruff KM, Strobel SA. Ligand binding by the tandem glycine riboswitch depends on aptamer dimerization but not double ligand occupancy. RNA. 2014;20:1775–88.

26. Babina AM, Lea NE, Meyer MM. In vivo behavior of the tandem glycine riboswitch in bacillus subtilis. mBio. 2017;8:1602.
27. Cochrane JC, Lipchock SV, Strobel SA. Structural investigation of the GlmS ribozyme bound to its catalytic cofactor. Chem Biol. 2007;14:97–105.
28. Collins JA, Irnov I, Baker S, Winkler WC. Mechanism of mRNA destabilization by the glmS ribozyme. Genes Dev. 2007;21:3356–68.
29. Klein DJ, Ferre-D'Amare AR. Structural basis of glmS ribozyme activation by glucosamine-6-phosphate. Science. 2006;313:1752–6.
30. Roth A, Nahvi A, Lee M, Jona I, Breaker RR. Characteristics of the glmS ribozyme suggest only structural roles for divalent metal ions. RNA. 2006;12:607–19.
31. Watson PY, Fedor MJ. The glmS riboswitch integrates signals from activating and inhibitory metabolites in vivo. Nat Struct Mol Biol. 2011;18:359–63.
32. McCown PJ, Winkler WC, Breaker RR. Mechanism and distribution of glmS ribozymes. Methods Mol Biol. 2012;848:113–29.
33. Xin Y, Hamelberg D. Deciphering the role of glucosamine-6-phosphate in the riboswitch action of glmS ribozyme. RNA. 2010;16:2455–63.
34. Lünse CE, Scott FJ, Suckling CJ, Mayer G. Novel TPP-riboswitch activators bypass metabolic enzyme dependency. Front Chem. 2014;2:53.
35. Strobel B, Spöring M, Klein H, et al. High-throughput identification of synthetic riboswitches by barcode-free amplicon-sequencing in human cells. Nat Commun. 2020;11:714.
36. Yokobayashi Y. Aptamer-based and aptazyme-based riboswitches in mammalian cells. Curr Opin Chem Biol. 2019;52:72–8.

Small Noncoding RNA, microRNA in Gene Regulation

8

Kristie Shirley, Kathryn Reichard, and Neena Grover

Contents

K. Shirley · K. Reichard · N. Grover (✉)
Department of Chemistry and Biochemistry, Colorado College, Colorado Springs, CO, USA
e-mail: k_shirley@coloradocollege.edu; Kathryn.reichard@coloradocollege.edu;
ngrover@ColoradoCollege.edu

© Springer Nature Switzerland AG 2022
N. Grover (ed.), *Fundamentals of RNA Structure and Function*, Learning Materials in
Biosciences, https://doi.org/10.1007/978-3-030-90214-8_8

Keywords

small non-coding RNA · miRNA lifecycle · miRNA biogenesis · miRNA regulation · miRNA cleavage

What You Will Learn
A large portion of the human genome is devoted to producing noncoding RNA. A majority of ncRNA are involved in complex regulatory networks that maintain essential cellular functions, including the immune response. In this chapter, we learn about the biochemical steps involved in generating microRNA (miRNA) via the canonical pathway. The role of the Microprocessor Complex, Dicer and Argonaute proteins in generating miRNA will be discussed. The biochemical processes involved in miRNA are similar to those utilized by other small ncRNA. Any disruption in the regulation of ncRNA pathways causes diseases. Understanding the complex network of regulatory interactions has important implication for RNA-based therapies.

Learning Objectives
After reading this chapter, the students should be able to

- Illustrate the key steps in canonical pathway of microRNA biogenesis.
- Compare the structures and mechanisms of RNase III enzymes involved in miRNA pathways to RNA cleavage by ribozymes.
- Explain the role of miRNA in the regulation of mRNA.
- Delineate the link between miRNA dysfunctions and disease.

8.1   Introduction

The coding regions of the human genome were sequenced to 99.99% accuracy in 2003. This new information added new layers to our understanding of the cellular processes [1]. The human genome is over three billion bases, with ~20,000 protein-coding genes. The number of genes discovered in the human genome is far fewer than was expected. For comparison, the rice genome is ~400 million bases with ~40,000 genes; wheat genome is ~16 billion bases with ~107,000 genes; a newt (salamander) genome is 10 billion bases with ~23,000 genes [2]. The genome of the bacteria *E. coli* has approximately 4.6 million bases with ~5000 genes. What does this information mean for the complexity of organisms? (Chap. 4, Spliceosome).

A high throughput transcriptomic analysis of human genome by the ENCODE (*Encyclopedia of DNA Elements*) Project shows a small percent (1–2%) of our genome is devoted to protein-coding regions but a surprisingly large percent (>80%) is transcribed into RNA [3, 4]. Organisms are spending a large amount of energy to generate non-coding RNA (ncRNA), pointing to its importance in cellular functions. A small fraction of ncRNA are abundant, constitutively expressed, and are involved in routine functions of the cell. These RNA are sometimes referred to as infrastructure or housekeeping RNA. These include the transfer RNA (tRNA), ribosomal (rRNA), small nuclear (snRNA), and small nucleolar RNA (snoRNA).

A majority of the ncRNA are regulatory and are involved in transcriptional and transcriptional gene regulation. The ncRNA are considered small or large based on a cut off of 200-nucleotide. The small noncoding RNA include, microRNA (miRNA), piwi-interacting RNA (piRNA), small interfering RNA (siRNA). Large non-coding RNA include, long non-coding RNA (lncRNA) and circular RNA (circ RNA) [5–11]. New ncRNA are being discovered including, enhancer associated RNA (eRNA) and promoter-associated RNA (PAR), among many others. In this chapter, we will primarily focus on biochemical processes involved in the production of miRNA and briefly discuss their role in regulation. The functions of a few others ncRNA are briefly presented at the end.

Most non-coding RNA are part of RNA-protein complexes (RNP) that perform essential functions in gene expression and in remodeling of the eukaryotic genome. The ncRNA regulate a broad spectrum of developmental and post-developmental processes. Any deletion in these gene regulatory pathways impacts the development of every organism examined.

The ncRNA-based regulation pathways are an ancient mechanism found in all domains of life—bacteria, archaea, and eukarya [5–15]. Most human protein transcripts are under selective pressure to retain their miRNA binding sites, indicating their importance. Most ncRNA are expressed at a low level, in a tissue- and development-specific manner; some are expressed in response to the environmental stimuli.

At its core, gene regulation by small ncRNA is about specifically creating and directing a small complementary nucleic acid to a particular mRNA target. The double-stranded RNA (dsRNA) are then recognized and acted upon by associated "effector" proteins which mediate the downstream functions, leading to a variety of outcomes for the mRNA.

The discovery of RNA regulatory networks broke the paradigm that genes were exclusively under the control of transcription factors, or that transcriptional control was the key to the complexity of organisms. Notably, genomic studies examining complex diseases, such as cancer, point to a majority of the disease-causing loci (haplotype blocks) existing outside the protein-coding regions [12].

The emergence of RNA-mediated gene silencing preceded multicellularity. The regulation of gene expression by microRNA likely evolved from an earlier RNA interference (RNAi) mechanism. RNA interference likely helped unicellular organisms to recognize invading nucleic acids from viruses, plasmids, or transposons. RNAi depends on cleavage as a mechanism for disabling foreign nucleic acids. The RNAi mechanism is likely to have evolved into an expanded miRNA repertoire. The miRNA-based gene regulation system is proposed to be linked to development of complexity in plant and animal kingdoms. In this chapter we will focus on the principles of RNA-recognition and regulation seen in microRNA (miRNA). The proteins that cleave the RNA—Drosha, Dicer, and Argonaut—are all involved in multiple pathways and utilize the same biochemical principles discussed here. Other ncRNA are briefly mentioned at the end of the chapter. Our understanding of gene regulation is rapidly evolving and will prove instrumental in developing next generation of medicine—whether it is in silencing defective genes or in fighting pathogenic organisms.

8.2   Overview of microRNA (miRNA)

MicroRNA are regulatory RNA that act as posttranscriptional repressors of gene expression and are essential feature of development in eukaryotes. The miRNA (*lin*-4 and *let*-7) were first identified via genetic screens as important for developmental timing in *C. elegans*. Since then, miRNA have been found across animal and plant species. The miRNA are short, 20–24 nucleotides, single-stranded RNA sequences that often base pair with the 3′UTR of their target mRNA to guide their translational repression, deadenylation or degradation [16]. The negative regulation occurs posttranscriptionally through association with proteins in an *R*NA-*i*nduced *s*ilencing *c*omplex (RISC). MicroRNA are expected to regulate ~50% of the transcribed genes. A given miRNA may regulate many mRNA; mRNA may in turn be regulated by many different miRNAs. Most miRNA are expressed with particular spatial and temporal specificity and may function in specialized cell types under varying conditions. All this makes examining the role of a particular miRNA challenging. In many cases the expression patterns and target mRNA by specific miRNA are conserved. The miRNA-dependent regulation of mRNA is a vast regulatory network and defects in these regulatory pathways lead to human diseases, including neurological disorders, cancer, and cardiovascular malfunction.

8.3   Canonical Pathway for miRNA Biogenesis

A large percent of miRNA are found within introns (~50%) and are often transcribed at the same time as their host gene using RNA polymerase II or III. Non-intragenic miRNA are transcribed independently from their own promoters within exons or even from intergenic regions.

Often multiple miRNA sequences are found together. When miRNA share a seed sequence, these clusters are called families and are transcribed together in one long transcript. The transcript is further processed into individual mature miRNA duplexes. Target complementarity with miRNA ensures specificity and the fate of the target RNA.

The biogenesis of miRNA begins in the nucleus with transcription. Long primary miRNA (pri-miRNA) containing hairpin structures are the substrate for the Microprocessor complex. The hairpin is excised by the Microprocessor Complex—Drosha and DGCR8 complex (also called Pasha, partner of Drosha), to produce 60–70 nucleotide precursor hairpin (pre-miRNA). Drosha is a RNAse III endonuclease. The pre-miRNA is exported out of the nucleus by Exportin 5/RanGTP pathway to the cytoplasm where pre-miRNA binds to Dicer, another RNase III nuclease. Dicer produces 21–24 nucleotide duplex miRNA. One strand of this miRNA is loaded onto Argonaute protein to form the silencing complex called miRISC (*mi*RNA *i*nduced *s*ilencing *c*omplex). At this stage, miRNA strand either binds to mRNA perfectly and is degraded or it binds imperfectly and is repressed (Fig. 8.1).

8.4 Pri-miRNA Processing by the Microprocessor Complex

The microRNA are derived from a 60–100 nucleotide pre-miRNA hairpin structure that is derived from a longer primary miRNA (pri-miRNA). A single pri-miRNA may contain several different miRNA. The pri-miRNA structure contains several hairpin loops. Each hairpin structure contains flanking sequences necessary for its processing.

The pri-miRNA is approximately a thousand bases capped and polyadenylated RNA that contains one or more characteristic long hairpin motifs. These hairpins are unique to pri-miRNA and are used for recognition by processing proteins. A pri-miRNA hairpin stem is ~33–39 nucleotides with loops ranging from ~3 to 23 nucleotides; these are approximately three helical turns, with an imperfectly base paired stem—bulge loops are likely at position ~5–9 and ~16–21 relative to the apical loop (or ~16–21 and ~28–32 relative to the base of the hairpin) (Fig. 8.2a). One terminal of this stem consists of single-stranded RNA flanking segments; the region where single-stranded RNA meets the double-stranded RNA stem is referred to as the single-stranded/double-stranded RNA junction (ss/dsRNA).

The Microprocessor complex cuts the pri-miRNA in the nucleus. The cleavage of pri-miRNAs is performed by Drosha, an RNase III ribonuclease [8, 18, 19]. It is assisted by DGCR8 (DiGeorge syndrome critical region gene 8), an RNA binding protein in the nucleus. One copy of Drosha and two copies of DGCR8 form the microprocessor complex that allows for correct orientation of RNA on Drosha. Both proteins contain double-stranded RNA binding domains. Drosha has a RNase III domain. DGCR8 heme-binding domain binds to the terminal loops of the pri-miRNA.

The local structures and length of the stems in RNA influence the cleavage efficiency and site selection by Drosha, influencing the population of miRNA produced. The stem region of the hairpin may contain unpaired regions and is divided into upper and lower stem based on Drosha cleavage sites (Fig. 8.2a). DGCR8 binds to the apical loop of RNA; the interactions with RNA are strengthened upon hemin binding to DGCR8. The DGCR8 interaction with the apical loop ensures that Drosha cuts exactly 22 nucleotides from the apical junction (where the apical loop meets the upper stem) and hence, DGCR8 plays an

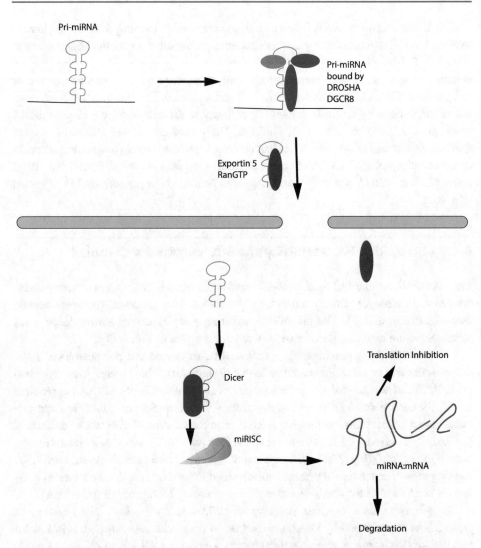

Fig. 8.1 The miRNA biogenesis via the Canonical Pathway. Long pre-miRNA containing hairpin are processed in the nucleus by the Microprocessor complex (Drosha-DGCR8). These hairpin structures are exported to the nucleus by Exportin5/RanGTP-dependent process. The hairpin RNA is processed by Dicer and one strand, miRNA, strand is loaded on to the RISC complex. The RISC complex-based interaction of miRNA with mRNA lead to degradation of mRNA or inhibition of translation

important role in choosing the miRNA sequence. At the other end of the stem-loop is the basal region, with 5′ (5p) and 3′ (3p) basal junction regions. Drosha cuts ~13 nucleotides from the basal junction region and recognizes specific RNA sequences. Multiple mismatches and wobble base pairs in upper stem regulate efficiency and accuracy of RNA cleavage by Drosha. Splicing factors, RNA editing based sequence changes, and

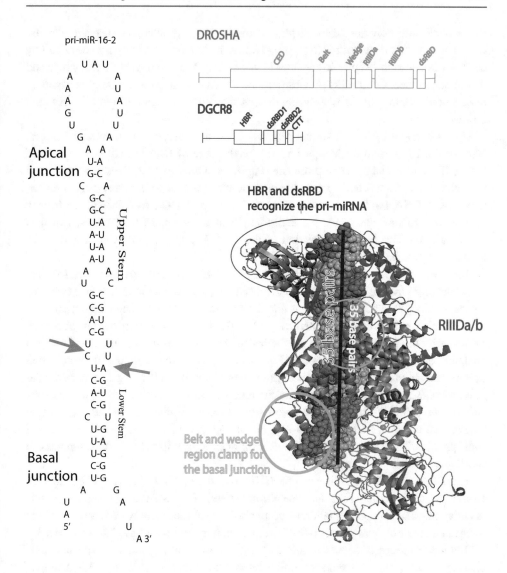

Fig. 8.2 The microprocessor complex. The pri-miRNA-16-2 is shown with apical and basal junctions marked. The cleavage sites are shown by the red arrows. Drosha is composed of a central domain (CED), RNase III domains (RIIIa/b), and double-stranded RNA binding domains. DGCR8 has heme-binding domain (HBR), dsRNA binding domains (dsRBD), and C-terminal tail (CTT). One copy of Drosha (green) and two copies of DGCR8 (blue and burnt orange) bind to the pri-miRNA (purple/orange sphere). Nearly 35 base pairs are held by Drosha (green) to the position of apical regions to interact with the DGCR8 (brown circle); the belt and wedge regions interact with the basal junction (light green circle). This serves as ruler to position the RNA correctly for cleavage by RNaseIII domains in Drosha (gray circle is in the area of RIIIa/b and dsRNA binding domain of Drosha) [17]. (Image created using PDB file 6V5B using PyMol)

single nucleotide polymorphism all play a role in selecting alternate cleavage sites by Drosha to generate different miRNA. miR-22 was the first miRNA shown to be edited by enzymes that deaminate specific adenines to inosines (ADAR1 and ADAR2) in human and mouse brain tissue. Other miRNA substrates have been shown to be processed in a manner that reduced cleavage by Drosha. In some cases, these changes cause differential targeting downstream.

Upon binding pri-miRNA, various regions of Drosha undergo conformational changes. Most of the RNA stem docks against the globular core of Drosha that is composed of RNase III domain and the central domain (Fig. 8.2b). Other parts of Drosha wrap around the RNA to make stable and specific interactions and in turn adopt a more rigid conformation. The dsRNA binding domains of both Drosha and DGCR8-1 interact to form a continuous structure that binds the entire length of the stem, 35 nucleotides, and thus acts as a ruler. The correct orientation of the dsRNA helps orient the correct sites for cleavage on Drosha.

Microprocessor Activity Modifiers. As many as 20 different proteins have been found to associate with Drosha, a few of them have been characterized as modifiers of Microprocessor activity [20–22]. Many proteins that are involved in mRNA processing or transcription initiation are involved in interacting with Drosha. Thus, which hairpin structures interact with the Microprocessor and whether their cleavage is up- or downregulated is a complex and well-regulated process. For example, the tumor suppressor protein p53 becomes active under stress. It binds to promoters of specific miRNA genes, in particular those in *miR-34* family, to up-regulate their transcription. In other cases, p53 recruits specific pri-miRNA to the Microprocessor, causing enhanced processing. These upregulated miRNA, *miR-16-1*, *miR-143*, and *miR-145*, target cell proliferation and cell cycle proteins. Thus, the regulation of miRNA biogenesis reinforces the tumor suppressive function of p53.

The DEAD-box RNA helicases p68 (DDX5) and p72 (DDX17) co-precipitate with Drosha and are important for select miRNA processing. Disruption of p68 and p72 results in embryonic and neonatal lethality, respectively. The helicase activity of these enzymes is implicated in rearrangement of pri-miRNA or dislodging inhibitory proteins from RNA.

The hairpin loops of some pri-miRNA play a key role in regulation of specific RNA binding proteins and can enhance or suppress Microprocessor activity. For example, hnRNP A1 has many roles in mRNA metabolism, including alternative splicing and its export from the nucleus. hnRNP A1 binds the terminal loops as well as the bottom stem of the hairpin to improve Microprocessor access for cleavage.

The structures of pri-miRNA play a regulatory role in the processing of the mature miRNA that reside within them. For example, *miR17~192* family is transcribed as a polycistronic transcript and is often seen overexpressed in several human cancers. The pri-miRNA forms a tertiary structure that changes the solvent accessibility of the hairpins within the 3' core and thus prevents Drosha from efficient processing of the interior of the transcript; this result in differential processing of the RNA, resulting in more miRNA produced from the exterior regions [23]. Not surprisingly, the *miR17~192* sequence is highly conserved. These conserved sequences serve to maintain a proper fold, effectively protecting the $3'$ core pri-miRNA from enzymatic processing.

8.5 RNase III Endonucleases Cut dsRNA

All RNase III proteins are Mg^{2+}-dependent endonucleases that act on double-stranded RNA and contain a characteristic ribonuclease (RNase III) domain [24–26]. These vary in length from ~200 to 2000 residues and have been subdivided into three classes. Class 1 RNase III enzymes are the simplest. These have a dsRNA binding domain (dsRBD) and a single nuclease domain. Class 2 enzymes have two RNaseIII domains, labeled as a and b. Class 3 enzymes are the largest, with a helicase domain (N-terminal DExD/H) in addition to the RNase III and dsRBD domains; these also contain a small DUF283 (domain of unknown function) and a PAZ domain.

Class I enzymes are ubiquitous in bacteria, bacteriophage and some fungi and function as homodimers, with a single combined active site—a catalytic valley, which contains two discrete dsRNA binding motifs (RBM). Amino acid side chains from RBD form hydrogen bonds with non-bridging phosphate oxygen atoms and $2'$-hydroxyl in the dsRNA backbone. Magnesium ions are a required part of the cleavage mechanism (discussed below).

Class 2 enzymes, which include Drosha, are involved in processing miRNA and rRNA. Drosha cuts pri-miRNA indiscriminately in absence of DGCR8 in humans. DGCR8 is involved in recognizing the hairpin and position 11 nucleotides away (1 turn) and forms a ruler for RNaseIII cleavage. DGCR8 contains a proline-binding WW domain that interacts with prolines on Drosha. Thus, other proline-binding proteins are likely to interact with Drosha and modify its specificity.

Class 3 enzymes include Dicer, an enzyme whose structure will be discussed in the following section.

RNase III enzymes have a dimeric catalytic domain structure that allows for cleavage of dsRNA. The cleavage sites have four acidic residues that are strictly conserved in RNase III proteins and stabilize the two magnesium ions involved in the reaction mechanism. A water molecule is activated by one magnesium ion to act as the nucleophile. The second magnesium facilitates the departure of the $3'$-oxygen (Fig. 8.3).

8.6 Moving Out of the Nucleus

The newly created pre-miRNA are exported from the nucleus to the cytoplasm by Exportin-5 in a RanGTP-dependent manner. Exportin 5 binds to substrates that have short $3'$ overhangs, terminal loops of at least four nucleotides. Exportin 5 (XPO5) binds to RNA with high affinity in a RanGTP-dependent manner. The hydrolysis of GTP in the cytoplasm causes dissociation and release of the RNA to Dicer. The short $3'$ overhang on

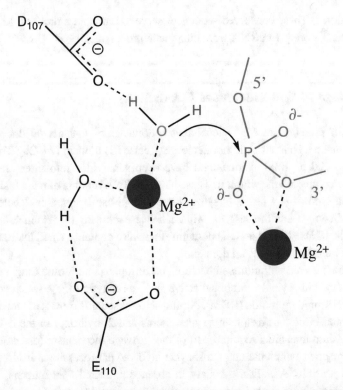

Fig. 8.3 The cleavage mechanism of RNase III nucleases. A water molecule is activated by one magnesium ion to act as a nucleophile on the phosphodiester linkage. The second magnesium stabilizes the transition state and the leaving group

the pre-miRNA are also necessary for processing by Dicer. Interestingly, Exportin-5 is also responsible for export of Dicer mRNA out of the nucleus.

8.7 Dicer Converts pre-miRNA into miRNA

Once in the cytoplasm, the pre-miRNA strands are processed by the RNase III endonuclease, Dicer [27–33]. Dicer cuts the hairpin-containing miRNA to short dsRNA and loads the correct single strand onto the Argonaute protein as part of the RISC complex.

Dicer is a 200-kilodalton multidomain elongated (100 Å long, 30–50 Å wide) protein that is used to process both miRNA and siRNA (discussed later). Dicer functions as a monomer using its two RNase III domains to form a dimeric catalytic site. Some organisms, like *H. sapiens and C. elegans* encode one copy of Dicer that processes different substrates. Other organisms, like *Drosophila melanogaster,* encode different Dicers with specialized functions [33]. Human Dicer does not hydrolyze ATP. Invertebrate Dicers hydrolyze ATP for their anti-viral functions.

Dicer contains the following domains: N-terminal DExD/H-box helicase domain, HELICc (helicase conserved carboxy-terminal domain), DUF283 (domain of unknown function), PAZ (Piwi/Argonaute/Zwille) domains, RNase IIIa and IIIb domains, and dsRNA binding domains. Dicer is an L-shaped protein with amino terminal helicase domains forming the clamp at base of the L. The PAZ and platform domain forms the head region. The RNase III domain, the dsRBD domains form the body. The PAZ and ribonuclease domains are connected by a linker helix that runs along the body—this connector helix sets the distance (ruler) for cleavage on the dsRNA. Cryo-EM structures of Dicer shows the expected L-shape (Fig. 8.4).

The PAZ domain which recognizes the 3' overhang on the RNA substrate and a phosphate binding pocket recognizes the phosphorylated 5'-end of the small RNA. This domain also contains extra basic residues that might affect the handing off the substrate to the next protein. Each catalytic RNase domain cuts one strand of the dsRNA. The DUF283 domain can bind single-stranded RNA.

In the canonical pathways, precursor miRNA (pre-miRNA) are cleaved by Dicer. Dicer produces small RNA that are typically 21–25 nucleotides long and have a 2-nucleotide overhang, a 5' phosphoryl and 3'-hydroxyl groups. It follows the same RNase III mechanism discussed in Sect. 8.5. The distance between the terminus binding PAZ domain and the RNase III domain determines the length of the cleavage product (Fig. 8.4). Active Dicer recognizes many types of dsRNA substrates for cleavage. The cleavage results in removal of the hairpin loop from the end of the pre-miRNA, leaving a short dsRNA duplex structure called miRNA/miRNA* (old nomenclature) or 5p/3p isoforms (new nomenclature).

8.8   Argonaute Proteins Are Part of the RNA-Induced Silencing Complex (RISC)

Dicer associates with Argonaute proteins to form various effector complexes, including the RNA-induced silencing complex (RISC). The duplexed 5p/3p is unwound to use one strand as a guide RNA (miRNA); the other strand, the passenger strand, is degraded. The loading of Argonaute proteins with the correct RNA strand from the dsRNA produced by Dicer entails recognition of the guide 5'-end by the mid domain and the 3'-end by the PAZ domain [15, 34–36]. The slicer active site resides in the PIWI domain. The strand with the thermodynamically less stable 5'-end becomes the guide RNA.

Since strand selection is relatively flexible, there are times when the passenger strand can be selected and loaded onto Argonaute as well. This process, called "arm-switching," has been noted in studies comparing miRNA isoforms between tissues. One particular example of this is miR142 where the 5p isoform is found in ovarian, testicular, and brain tissues while the 3p isoform is more common in embryonic tissue samples.

Argonaute proteins are found in eukarya, archaea and bacteria. Although there is very little sequence homology between bacterial and human Argonaute proteins, their overall architecture and function are remarkably conserved. They contain an N-terminal domain, a

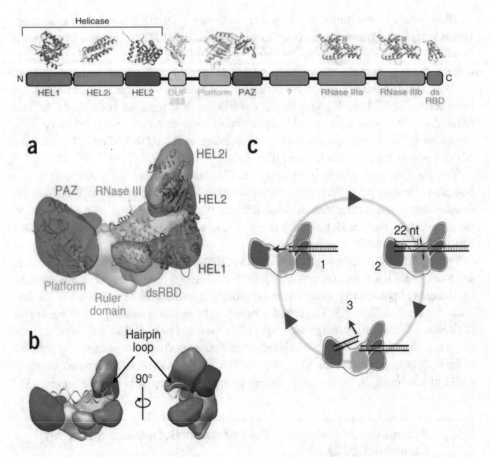

Fig. 8.4 Cryo-EM structure of dicer. The domain organization of Dicer is shown with structures of individual domains shown above. (**a**) A segmented map of human Dicer protein with various domains of Dicer docked in it. (**b**) A model for pre-miRNA recognition; the stem-loop is modeled into the cleft of the helicase. (**c**) A model for processive "dicing": step 1, The nuclease domain translocates the RNA into the nuclease core; step 2, the PAZ domain recognizes dsRNA end to position it for cleavage by the RNase domain (orange); step 3, the product siRNA is released. (Figures from [31])

PAZ domain, a middle (MID), and an RNase H like Piwi domains. Argonaute proteins are bilobed; the lobes are connected by two linkers, L1 and L2 (Fig. 8.5). The N-terminal lobe has the N and PAZ domains connected by the L1 linker. The C-terminal lobe has the MID and PIWI domains connected by L2 linker; the L2 linker forms the base of the central cleft containing the active site. The cleft is larger in the eukaryotic Argonautes than the prokaryotic counter parts, which may have implication for the target binding. All domains of Argonautes contact the guide RNA. The 5'-phosphate of the guide RNA is buried in a hydrophilic pocket of the MID domain. A rigid loop, called the nucleotide specificity loop, checks the specificity of the 5'-base. The 3'-end of the guide RNA uses its sugar 2'- and

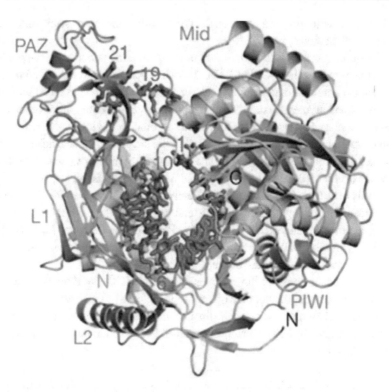

Fig. 8.5 Structure of argonaute protein bound to guide DNA. Individual domains of Argonaute protein from *T. thermophilus* are color coded. The guide DNA strand is shown as red sticks with phosphate atoms in yellow. (Figure from [35])

$3'$-hydroxyls to form hydrogen bonds, along with the sugar-phosphate backbone of the two nucleotides preceding the terminal nucleotide, in a shallow pocket of the PAZ domain, hence, no check for base identity occurs here. The nucleotides 2–7 are termed the "seed region" that binds to the target. Extensive hydrogen bonding to the phosphate backbone and the salt bridges orients the guide RNA in the A-form conformation, preorganized and ready to bind the target. Nucleotide 2–6 are solvent exposed and hence accessible for forming canonical base pairs with the target RNA. The rest of the RNA is threaded through the hydrophilic channel in the center of the protein making extensive hydrogen bonds.

Loading of RNA on Argonaute proteins is more efficient with duplexed RNA than with single-stranded guide RNA, with a preference for $5'$-uracil or adenine on the guide strand in humans. This preference varies between different Argonaute proteins [34]. The position 2 nucleotide stacks with tyrosine. This amino acid is conserved as either tyrosine or threonine; this might represent a conserved hydrophobic interaction that facilitates the flipping out of nucleotide 1. It likely prevents nucleotide 1 from stacking on nucleotide in position 2. Bases in position 3 and 4 don't make contact with the protein but are part of a tilted A-form helical structure that form a continuous seed sequence. When the target nucleotides pair with position 2 through 4, it un-tilts the seed stack and facilitates repositioning of the helix.

The N-terminal domain is important for duplex unwinding and loading the guide RNA. The PAZ domain binds the $3'$-end of RNA. The MID domain binds the $5'$ end of the RNA via base stacking and ionic interactions and is further aided by bound metal ions or lysine residues. In the eukaryotes, guide RNA also recruits an arginine into the $5'$ binding pocket. The bases of the seed nucleotide are now ready to bind to the target RNA. Remarkably, Argonaute proteins are ten-times faster at binding the target RNA than a naked guide RNA might anneal to its target in free solution. The ability of Argonaute proteins to bind any seed sequence means that RISC can be programmed to silence any target sequence.

Hydrolysis of ATP is important for RISC activity. Several steps in RISC formation are ATP-dependent, including duplex loading and passenger strand removal. The duplex unwinding is not ATP-dependent. This led to a "rubber-band" model for loading RISC, where ATP hydrolysis pulls Argonaute into a stretched state. Its release from this state provides the energy for unwinding during RISC loading.

The PIWI domain contains the residues DDX/DEDX (X = E or H) that form the cleavage site and contain the bound Mg^{2+} ions. The cleavage mechanism is similar to the two-metal mechanism discussed above (for Drosha and Dicer) where an activated water molecule is the nucleophile. The water molecule and magnesium ions are held by the same Asp-Asp-His residues.

8.9   miRNA-Mediated Gene Silencing

Argonaute proteins lightly bind and scan the target mRNA. Upon interactions with the target, cleavage of the mRNA occurs. The cleavage site is on the mRNA, on the strand opposite nucleotides 10 and 11 on the guide RNA, from the anchored $5'$-end of the gRNA. This sequence specific target cleavage is also called "slicing." In case of target non-complementarity, an alternate outcome is gene repression, leading to the mRNA being sequestered into P-bodies (i.e., translational silencing). A third method of target silencing is transcriptional silencing (TGS), which involves regulating the heterochromatin.

The RISC complex binds to complementary mRNA sequences in $3'$ UTRs to either suppresses their translation or to degrade the mRNA using the endonuclease domain of the Argonaute proteins. However, genomic analyses indicate a greater complexity of the silencing response [35–40]. Another interesting area that is not discussed here is the pathogen-derived miRNA which modulate the host cellular systems.

Target Specificity. Specificity for target recognition is mediated, in a large part, by the $5'$ 2–7 nucleotide seed sequences within miRNA which are highly conserved within miRNA families. The seed sequence is complementary to the miRNA response elements (MREs) found within the target mRNAs, typically in the $3'$UTRs. The degree of complementarity between the miRNA seed sequence and MRE affects the way the mRNA is processed [35–

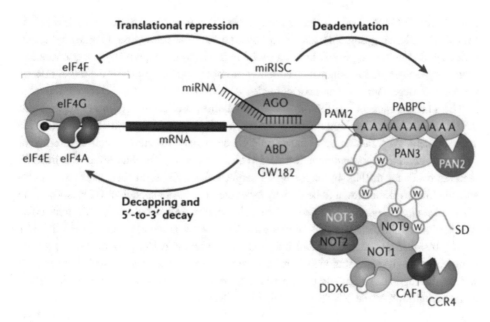

Fig. 8.6 Overview of the miRNA-mediated gene silencing. The miRNA loaded RISC complexes base pair with the mRNA, primarily in the 3'UTR. The Argonaute proteins interact with GW182 using the Argonaute binding domain (ABD) and a silencing domain; GW182 interacts with poly-A binding protein (PABPC) and the cytoplasmic deadenylase complex, PAN2-PAN3 and CCR4-NOT complexes to deadenylate the mRNA which are degraded by XRN1 in 5'-to-3' manner (not shown). The repression of translation by miRNA is expected to occur by inhibiting translation initiation by interactions with eIF4F complex. This complex is composed of eIF4E—a cap-binding protein (cap shown as a black filled circle), eIF4G—an adapter protein, and eIF4A—a DEAD-box helicase. (Figure from [37])

40]. Fully complementary mRNA sequences are usually degraded by Argonaute protein's endonuclease activity, but this interaction also destabilizes the 3' end of the miRNA which leads to its degradation as well. In animals, most interactions are not fully complementary due to mismatches in the central complementary region between the seed and MRE sequences. These mismatches allow for a looser binding and thus prevent endonuclease activity (Fig. 8.6).

miRNA-mediated Gene Regulation. While important for target recognition, miRNAs do not repress the target themselves. The Argonaute proteins bind various different protein factors, including helicases, chromodomain proteins, methyltransferases, RNA polymerases, and hence, they recruit various other proteins to the mRNA to allow a divergent-silencing response.

When Argonaute and its associated proteins are artificially tethered to the 3'UTR of an mRNA, translation is halted even in the absence of miRNA-MRE interactions. Argonaute proteins work together with glycine-tryptophan repeat containing protein (GW182) to facilitate a platform for the binding of various regulatory proteins; blocking this interaction disrupts miRISC-mediated silencing completely. The C-terminal domain of GW182 is named the 'silencing domain,' and is responsible for gene repression. GW182 represses genes by binding to the poly-A binding proteins (PABPs) of target mRNA sequences and promoting deadenylation which subsequently promotes the decapping of the target mRNA and its 5'-3' degradation by an exoribonuclease1 (XRN1).

The complexities of miRNA-mediated regulation are becoming more apparent as various miRNA-mediated regulatory pathways and the regulation of miRNAs themselves are examined. Any factors that change mRNA secondary structures, such as alternative splicing, alternative poly-adenylation of 3'UTRs, or editing can thus affect the miRNA regulation of that mRNA by exposing or sequestering MREs. In fact, alternative poly-adenylation has been shown to be critical for expressing tissue-specific 3'UTR isoforms in *Caenorhabditis elegans*. Alternative poly-adenylation relies on using different poly-adenylation signal elements (PAS); these are hexameric sequences found in the 3'UTRs of mRNAs that are often regulated by miRNAs. Differential isoforms of the 3'UTR are more common in genes that are expressed across many tissues and as such may require differential regulation. This tissue-specific regulation is accomplished through the differential recognition of the 3'UTR by regulatory miRNAs.

8.10 miRNA Regulation

The miRNA are regulatory molecules that have to be regulated [38–40]. Transcription of pri-miRNA is under the control of transcription factors and enhancers. Posttranscription regulation can occur via modulating the activity or levels of microprocessor complex or by controlling levels of individual miRNA via interactions with RNA binding proteins. miRNA biogenesis rate in *Drosophila* is 17 to >200 molecules/min/cell. The loading in Argonaute proteins to produce miRISC is significantly slower, thus, creating a kinetic bottle neck. Nearly 40% of miRNA produced are expected to get degraded before they are loaded. Thus, simple expression-level of miRNA may not be sufficient to understand its effectiveness. In general, miRNA are considered to be abundant and stable; however, individual miRNA have varying stability from minutes to days; neuronal miRNA have high turnover which is linked to neuronal activity. Effective repression by miRNA requires high concentration of miRISC relative to the target. To achieve this, syntheses of many miRNA occur ahead of their functional need in the cell. Circular RNA, discussed below, also play a role in miRNA regulation.

8.11 Circular RNA

RNA circles are covalently closed circles that were first identified in viroids in 1970s and then seen in electron micrographs of cytoplasmic fractions in eukaryotes; in eukaroytes, circRNA were considered "junk" by-products of splicing [11, 41–43]. Now many thousands of circular RNAs (circRNAs) have been identified as important regulators of cellular functions. Humans have over 100,000 unique circRNA, a number fivefold greater than the number of genes that code for proteins.

CircRNA are produced by non-canonical splicing event called back splicing. A downstream splice-donor site is covalently linked to an upstream splice-acceptor site to create a circular RNA. CircRNA show cell- and tissue-specific expression patterns whose biogenesis is regulated by cis-acting elements and trans-acting factors. Some of circRNAs are abundant and regulate miRNA expression by regulating protein functions or by being translated themselves.

CircRNAs are difficult to detect due to a lack of a 5′ or 3′ end. They are incredibly stable and are highly conserved. Despite lack of poly(A) tail, circRNA localize to the cytoplasm using ATP-dependent RNA helicase DDX39A (also known as RNA helicase URH49) and spliceosomal RNA helicase DDX39B (also known as DEAD-box protein UAP56), with different specificities for different length of circRNA (>1200 nucleotides for UAP56 and <400 for URH49).

Mechanisms of circRNA biogenesis are likely dependent on the canonical splicing pathways. Depleting the number of splicing factors in *Drosophila* caused an increase in the production of circRNA. A-to-I editing by ADAR enzymes suppressed the biogenesis of circRNA that rely on base pairing between inverted repeats. Epigenetic changes in histones may directly affect circRNA production by causing changes in alternate splicing of certain RNA. Transcription elongation rate is higher for circRNA-producing genes as compared to non-circRNA producing genes.

Endogenous circRNA are highly expressed in most human tissues. Most circRNA are produced from protein-coding genes and contain one or more exons. A high number of circRNA are upregulated during neurogenesis and downregulated in cancer cells. CircRNA are expected to play a role in cancers, diabetes mellitus, cardiovascular diseases, chronic inflammation, and neurological disorders. The presence or absence of these RNA is expected to play a role in diagnosing diseases.

The third position of the codon is generally not highly conserved but is found to be more conserved in circRNA, suggesting that these RNA play an important non-coding role that may be conserved. CircRNA are expected to act as sponges or decoys for miRNA protecting target mRNA. CircRNA containing RNA binding protein motifs may function as sponges or decoys for proteins to regulate their functions. A few circRNA function as scaffolds for proteins to facilitate colocalization of enzymes to influence the rate of reactions. Under certain conditions, those circRNA that contain internal ribosome entry site and a start site get translated to make peptides.

CircRNA harbor MRE (miRNA regulatory element) sequences which compete with the host gene for miRNA binding and act as "miRNA sponges". Once bound, the target miRNA is either degraded or stabilized. This regulatory form of circRNA are a perfect example of competitive endogenous RNA (ceRNA) which are defined by the use of shared MRE sequences to regulate miRNAs. circRNA's competitive binding for miRNAs is a potent form of regulation since circRNA are not easily broken down they can effectively stabilize miRNA or degrade them without having to constantly be replenished. circRNA can also regulate miRNAs indirectly through their regulation of alternative splicing. miRNA-based regulation is particularly sensitive to alternative splicing.

8.12 Small Interfering RNA

Small interfering RNA, siRNA, are 20–25 nucleotide double-stranded regulatory molecule [39, 44–46]. The siRNA response was first characterized in transgenic plants which have DNA from another species. These plants demonstrated a unique response to RNA transcribed from the introduced DNA. If the introduced genes had sequence complementarity to genes in the native species, the native mRNA was decreased in concentration. This response indicated that the sequence complementarity between the introduced mRNAs and the native mRNA was somehow linked to the knockdown effect observed in the plants. The transgenes create double-stranded RNAs that recognize complementary sequences (in native mRNA) which marks these mRNAs for degradation.

Dicer cleaves the dsRNA into siRNA. Dicer cleavage of long dsRNA into siRNA is very similar to pre-miRNA cleavage into miRNA. Because the hairpin structure of pre-miRNAs is not present on dsRNA precursors to siRNA, the molecular ruler mechanism is not as precise, and the exact siRNA created by Dicer cleavage is not the same size for every molecule. As with miRNA, the cleaved siRNA is then loaded onto an AGO protein and mediates mRNA degradation through the RISC.

siRNA can be either endogenous or exogenous. Both exo- and endo-siRNAs mediate mRNA degradation in a manner similar to miRNA. Exogenous siRNAs are produced from viral dsRNAs, transcribed transgenic genes, or man-made dsRNAs. Endogenous siRNAs, endo-siRNAs, are generated in plants through direct transcription and transcription of inverted repeats of transposons. *C. elegans* produce a class of endo-siRNA called tiny-non-coding RNA (tncRNA) via a Dicer-dependent pathway. Mammalian endo-siRNA derives from transposons and hairpin precursors. Their synthesis is Dicer dependent; they are implicated in silencing of transposable elements.

8.13 Piwi RNA

Piwi RNA (piRNA) are 28–33 nucleotides, single-stranded small regulatory RNA that associate with an Argonaute family protein named Piwi (P-element induced wimpy testis) [39, 47, 48]. As the name implies, Piwi proteins are a germline enriched molecule necessary for germline-specific events in meiosis, including cell maintenance and cell differentiation. piRNAs have 85–95% conservation of a 5′ uridine, have a single-stranded primary transcript, and their biogenesis is Dicer independent.

Tens of thousands of diverse, mature piRNAs emerge from only 50–100 defined primary piRNA sequences. Thus far, mature piRNA processing from the restricted primary transcripts is believed to be stochastic. This belief is reinforced by a very limited amount of piRNA conservation between species. However, sequences of the most highly conserved piRNAs are conserved, indicating that a not yet discovered pathway may regulate processing.

Despite having limited sequence conservation, piRNAs have very strong loci conservation and emerge from a few specific regions of homology on mammalian chromosomes. piRNAs seem to regulate transposon silencing, which may be part of their role in germ cell differentiation. Mechanistic details of piRNA are being further investigated.

8.14 Other Small Regulatory RNA

The repeat associated small interfering RNAs (rasiRNAs) were found in Drosophila and are not known to exist in any other species. rasiRNAs are a 28 nt, testes enriched molecule and are transcribed from retrotransposon repeat sequences [39]. These molecules have a $2'$-O-methyl on the ribose on their $3'$ end. Piwi family members in Drosophila associate with rasiRNAs. Originally, piRNA was isolated in an attempt to locate rasiRNA in Mammalia. Despite their similar Piwi association and testes enrichment, piRNA genetic loci are not associated with repeat-rich regions. rasiRNA and piRNA are not considered to be in the same class because of their divergent biogenesis.

Like rasiRNAs, there are other small RNAs that associate with the RISC to regulate cellular mRNA in certain species. C. elegans express a small RNA in concentrations ten-fold less than miRNA called 21U-RNAs based on their conserved 21st uridine residue. 21U-RNAs originate from chromosome IV downstream from a conserved regulatory sequence element. Although their function is unknown, conserved regulation of the molecule's transcription indicates they have conserved cellular function. Another small 26–30 nt length RNA class, scanRNAs are present in protozoans and their biogenesis is Dicer dependent. This class of small regulatory RNA participates in chromatin modification and contributes to DNA elimination.

8.15 Long Noncoding RNA (lncRNA)

Thousands of long non-protein-coding RNA, longer than 200 nucleotides, are made in human cells. These are primarily transcribed by polymerase II (Pol II). The lncRNA derived from long intergenic regions (lincRNAs) are several kilobases to 200 kb molecules; these are transcribed from regions lacking open reading frames. Mature lncRNA have a $5'$ cap and poly (A) tail like the mature mRNA and are spliced similarly to mRNA. These long molecules were overlooked for a longtime because they resemble mRNA and were considered to be "junk transcription" with little to no functional relevance. In the last two decades, lncRNA were shown to be tissue- and condition-specific and were found to regulate a wide-range of cellular processes including, DNA replication and DNA repair, membrane-less nuclear bodies, stability and translation of cytoplasmic mRNA, and signaling pathways. Thus, their importance in neuronal disorders, immune response and cancer is emerging as an important story in cellular regulatory pathways.

Several lncRNAs control adjacent genes (cis-acting) by affecting transcription or chromatin remodeling. Two genes, HOTAIR and HOTTIP, function in cis to regulate HOX loci. Hox genes are necessary for defining body segmentation early in animal development. HOTTIP is abundant in distal segments and alters methylation patterns to turn on Hox genes that generate distal embryo development. HOTAIR regulates the chromatin state of the HOXC locus and is an important for skin development.

Some lncRNA work away from their loci; these trans-acting lncRNAs are implicated in maintaining the pluripotent state. Nanog and Oct4 are two transcription factors present in embryonic stem cells that indicate pluripotency. The concentration of these transcription factors decreased when certain lncRNA were knocked out. The exact mechanisms of lncRNA chromatin modification are being examined, as many co-immunoprecipitate with at least one chromatin associating protein. This indicates that the lncRNA may function structurally, catalytically or as a guide in larger protein-based chromatin modifying complexes.

8.16   Antisense RNA

Antisense RNA is a general class of ncRNA named for a diverse array of molecules that originate from the antisense strand of protein-coding genes. These RNAs are produced because of bi-directional expression; their sequences have reverse complementarity to the protein-coding sense transcript. Although complementary sequences are used by the RISC to induce mRNA degradation, transcription of antisense transcripts does not lead to a decrease in the protein coded by the sense mRNA. This indicates that, contrary to original hypotheses, antisense RNAs do not directly negatively regulate their sense complement.

Most antisense RNA genes are imprinted reciprocally from the sense, coding strand. These antisense transcripts are expressed with developmental specificity and, because of their imprinting, may be a part of activation and inactivation of their sense partner. Currently, antisense RNAs seem to act directly on the genome, in cis to their sequence and the sense, coding strand.

8.17   ncRNA Research Methods

Noncoding RNAs are regulatory molecules, therefore, they are less abundant than mRNAs [49]. The small size of many regulatory molecules further complicates their isolation. Continuing to reassess total cellular RNA populations at various points in development and in various cell types provides information about the diversity of ncRNA in eukaryotic systems. Expanding RNA research requires parallel growth of research methods.

RNAomics is the term for genomic analyses for RNA expression. New in silico screens are being created to use software that models structure of putative noncoding RNAs. This software helps to identify RNAs whose identity is based on conserved structural components between species, but not conserved sequence.

New methods exploit known RNA–protein interactions to isolate non-coding RNAs that were not previously found using microarray and cDNA techniques developed to determine cellular transcription.

8.18 Implications in Human Diseases

A variety of ncRNA have been identified and are seen to play important role in cellular regulation; thus, dysregulation in these can cause disease [50–53]. About 20% of analyzed microRNAs map to fragile sites, chromosomal regions that are more susceptible to genetic alterations in human tumors. Many miRNAs play a role in either apoptosis or cell proliferation, mechanisms that, if dysregulated, can lead to oncogenesis. For example, human lung cancer demonstrates reduced expression of let-7 microRNA that correlates with disease pathology. Another example, Burkitt's lymphoma, alternatively demonstrates 100-fold upregulation of pre-miRNA-155. The dysregulation of microRNA could potentially serve as a biomarker to screen for cancer non-invasively and pharmaceutical regulation of miRNAs may have the potential to clinically alter the disease state.

MicroRNAs are also upregulated in the nervous system and dysregulated in multiple neurological conditions. Fragile-X Mental Retardation, several forms of Spinal Muscular Atrophy, traumatic brain injury, and Schizophrenia all have non-wild-type miRNA distribution. Because lncRNAs and antisense RNAs are involved in epigenetic alterations to the genome, improving understanding of their regulation will undoubtedly reveal connections to disease and therapeutic potential of these ncRNAs.

Some ncRNA have been implicated in cancers so often that they have been categorized as canonical tumor suppressors or tumor suppressor suppressors in which case they are called OncomiRs. Much effort is focused on using ncRNA expression profiles as a diagnostic marker to identify the location and tissue type of tumors.

Various ncRNA play diverse roles in regulating eukaryotic gene expression. Evolution based on ncRNA may well be a mechanism to increase nervous system and physiological complexity. Much has been learned about biogenesis and regulatory pathways of some ncRNA and a great deal more remains unexplored, some as a limitation of technology and some due to novel and unexpected functions not yet ascribed to RNA.

Take Home Message

- The canonical pathway of miRNA biogenesis starts with transcription of hnRNA.
- The primary miRNA is processed by the Microprocessor complex to produce hairpin structures that are transported out of the nucleus.
- Dicer acts on hairpin RNA to produce mature miRNA that are loaded into RISC complex.
- Argonaute proteins, in concert with other effector proteins, bind small RNA to their mRNA targets which results in different silencing pathways.
- Many different short and long ncRNA exist in the cell and are themselves regulated.
- Defects in ncRNA regulatory networks leads to dysfunction and disease.

References

1. International Human Genome Sequencing Consortium. Finishing the euchromatic sequence of the human genome. Nature. 2004;431:931–45.
2. Touchman J. Comparative genomics. Nat Educ Knowl. 2010;3:13.
3. Davis CA, et al. The encyclopedia of DNA elements (ENCODE): data portal update. Nucleic Acids Res. 2018;46:D794–801.
4. Pennisi E. ENCODE project writes eulogy for junk DNA. Science. 2012;6099:1159.
5. Morris K, Mattick J. The rise of regulatory RNA. Nat Rev Genet. 2014;15:423–37.
6. Mattick JS, Makunin IV. Non-coding RNA. Hum Mol Genet. 2006;15:R17–29.
7. Dexheimer PJ, Luisa C. MicroRNAs: from mechanism to organism. Front Cell Dev Biol. 2020;8:409.
8. Bartel DP. MicroRNAs: target recognition and regulatory functions. Cell. 2009;2:215–33.
9. Cech TR, Steitz JA. The noncoding RNA revolution-trashing old rules to forge new ones. Cell. 2014;157:77–94.
10. Storz G, Altuvia S, Wassarman KM. An abundance of RNA regulators. Annu Rev Biochem. 2005;74:199–217.
11. Li X, Yang L, Chen L. The biogenesis, functions, and challenges of circular RNAs. Mol Cell. 2018;3:428–42.
12. Freedman ML, et al. Principles for the post-GWAS functional characterization of cancer risk loci. Nat Genet. 2011;43:513–8.
13. Gottesman S. Micros for microbes: non-coding regulatory RNAs in bacteria. Trends Genet. 2005;21:399–404.
14. Sherafatian M, Mowla SJ. The origins and evolutionary history of human non-coding RNA regulatory networks. J Bioinforma Comput Biol. 2017;15:1750005.
15. Tolia NH, Joshua-Tor L. Slicer and the argonautes. Nat Chem Biol. 2007;3:36–43.
16. O'Brien J, Hayder H, Zayed Y, Peng C. Overview of MicroRNA biogenesis, mechanisms of actions, and circulation. Front Endocrinol. 2018;9:402.
17. Partin AC, Zhang K, Jeong C, et al. Cryo-EM structures of human Drosha and DGCR8 in complex with primary MicroRNA. Mol Cell. 2020;78:411–22.
18. Han J, Lee Y, Yeom KH, et al. Molecular basis for the recognition of primary microRNAs by the Drosha-DGCR8 complex. Cell. 2006;125:887–901.

19. Li S, Nguyen TD, Nguyen TL, et al. Mismatched and wobble base pairs govern primary microRNA processing by human microprocessor. Nat Commun. 2020;11:1926.
20. Finnegan EF, Pasquinelli AE. MicroRNA biogenesis: regulating the regulators. Crit Rev Biochem Mol Biol. 2013;48:51–68. https://doi.org/10.3109/10409238.2012.738643.
21. Farazi TA, Hoell JI, Morozov P, Tuschl T. MicroRNAs in human cancer. In: Schmitz U, Wolkenhauer O, Vera J, editors. MicroRNA cancer regulation: advanced concepts, bioinformatics and systems biology tools. Dordrecht: Springer; 2013. p. 1–20.
22. Gregory RI, Yan KP, Amuthan G, et al. The microprocessor complex mediates the genesis of microRNAs. Nature. 2004;432:235–40.
23. Chaulk SG, Thede GL, Kent OA, et al. Role of pri-miRNA tertiary structure in miR-17~92 miRNA biogenesis. RNA Biol. 2011;8:1105–14.
24. Court DL, Gan J, Liang YH, et al. RNase III: genetics and function; structure and mechanism. Annu Rev Genet. 2013;47:405–31. https://doi.org/10.1146/annurev-genet-110711-155618.
25. Nicholson AW. Ribonuclease III mechanisms of double-stranded RNA cleavage. WIREs RNA. 2014;5:13–48.
26. MacRae IJ, Doudna JA. Ribonuclease revisited: structural insights into ribonuclease III family enzymes. Curr Opin Struct Biol. 2007;17:138–45.
27. Song MS, Rossi JJ. Molecular mechanisms of dicer: endonuclease and enzymatic activity. Biochem J. 2017;474:1603–18.
28. Starega-Roslan J, Koscianska E, Kozlowski P, Krzyzosiak WJ. The role of the precursor structure in the biogenesis of microRNA. Cell Mol Life Sci. 2011;68:2859–71.
29. Park JE, Heo I, Tian Y, Simanshu DK, Chang H, Jee D, Patel DJ, Kim VN. Dicer recognizes the 5′ end of RNA for efficient and accurate processing. Nature. 2011;475:201–5.
30. Fabian MR, Sonenberg N. The mechanics of miRNA-mediated gene silencing: a look under the hood of miRISC. Nat Struct Mol Biol. 2012;6:586–93.
31. Lau PW, Guiley K, De N, et al. The molecular architecture of human dicer. Nat Struct Mol Biol. 2012;19:436–40.
32. Wilson RC, Tambe A, Kidwell MA, et al. Diver-TRBP complex formation ensures accurate mammalian MicroRNA biogenesis. Mol Cell. 2015;57:397–407.
33. Sinha NK, Iwasa J, Shen PS, Bass BL. Dicer uses distinct modules for recognizing dsRNA termini. Science. 2018;359:329–34.
34. Schirle NT, MacRae IJ. Chapter 4 – Structure and mechanism of argonaute proteins. In: Guo F, Tamanoi F, editors. The enzymes, vol. 32. Academic Press; 2012. p. 83–100. ISSN 1874-6047, ISBN 9780124047419. https://doi.org/10.1016/B978-0-12-404741-9.00004-0.
35. Wang Y, Sheng G, Juranek S, et al. Structure of the guide-strand-containing argonaute silencing complex. Nature. 2008;456:209–13.
36. Hundley HA, Bass BL. ADAR editing in double-stranded UTRs and other noncoding RNA sequences. Trends Biochem Sci. 2010;35:377–83.
37. Jonas S, Izaurralde E. Towards a molecular understanding of microRNA-mediated gene silencing. Nat Rev Genet. 2015;16:421–33.
38. Kai ZS, Pasquinelli AE. MicroRNA assassins: factors that regulate the disappearance of miRNAs. Nat Struct Mol Biol. 2010;17:5–10.
39. Farazi TA, Juranek SA, Tuschl T. The growing catalog of small RNAs and their association with distinct Argonaute/Piwi family members. Development. 2008;135:1201–14.
40. Hammond S. An overview of miRNA. Adv Drug Deliv Rev. 2016;87:3–14.
41. Barrett SP, Wang PL, Salzman J. Circular RNA biogenesis can proceed through an exon-containing lariat precursor. eLife. 2015;4:e07540.
42. Qu S, Yang X, Li X, et al. Circular RNA: a new star of noncoding RNAs. Cancer Lett. 2015;2: 141–8.

43. Barrett SP, Salzman J. Circular RNAs: analysis, expression and potential functions. Development. 2016;11:1838.
44. Bernstein E, Caudy AA, Hammond SM, Hannon GJ. Role for a bidentate ribonuclease in the initiation step of RNA interference. Nature. 2001;409:363–6.
45. Carthew RW, Sontheimer EJ. Origins and mechanisms of miRNAs and siRNAs. Cell. 2009;136: 642–55.
46. Naganuma M, Tadakuma H, Tomari Y. Single-molecule analysis of processive double-stranded RNA cleavage by drosophila Dicer-2. Nat Commun. 2021;12:4268.
47. Iwasaki YW, Siomi MC, Siomi H. PIWI-interacting RNA: its biogenesis and functions. Annu Rev Biochem. 2015;1:405–33.
48. Girard A, Sachidanandam R, Hannon GJ, Carmell MA. A germline-specific class of small RNAs binds mammalian Piwi proteins. Nature. 2006;442:199–202.
49. Hüttenhofer A, Vogel J. Experimental approaches to identify non-coding RNAs. Nucleic Acids Res. 2006;34:635–46.
50. Xu Z, YanY ZS, Dai S, et al. Circular RNAs: clinical relevance in cancer. Oncotarget. 2018;9: 1444–60.
51. Romero-Cordoba S, Salido-Guadarrama I, Rodriguez-Dorantes M, Hidalgo-Miranda A. miRNA biogenesis: biological impact in the development of cancer. Cancer Biol Ther. 2014;11:1444–55.
52. Ofek P, Tiram G, Satchi-Fainaro R. Angiogenesis regulation by nanocarriers bearing RNA interference. Adv Drug Deliv Rev. 2017;119:3–19.
53. Pudova EA, Krasnov GS, Nyushko KM, et al. miRNAs expression signature potentially associated with lymphatic dissemination in locally advanced prostate cancer. BMC Med Genet. 2021;13:129.

CRISPR-Cas Systems: The Science and Ethics of Gene Manipulation

9

Julia Poje and Neena Grover

Contents

Keywords

CRISPR · CRISPR ethics · CRISPR science · CRISPR cleavage · CRISPR gene editing

What You Will Learn

The CRISPR-Cas immune system in bacteria and archaea capture small pieces of viral DNA to incorporate into their own genome. Upon transcription into RNA, these

(continued)

J. Poje · N. Grover (✉)
Department of Chemistry and Biochemistry, Colorado College, Colorado Springs, CO, USA
e-mail: julia.poje@coloradocollege.edu; ngrover@ColoradoCollege.edu

© Springer Nature Switzerland AG 2022
N. Grover (ed.), *Fundamentals of RNA Structure and Function*, Learning Materials in Biosciences, https://doi.org/10.1007/978-3-030-90214-8_9

sequences bind with Cas proteins. The Cas-bound RNA is used for surveillance in the cell. When the same virus reinfects the host, the RNA finds its complementary sequence to form a RNA:DNA hybrid. Cas proteins then cut the viral DNA in the RNA:DNA hybrid. Different organisms utilize different Cas proteins. In this chapter, we will learn the general principles of CRISPR-Cas-based immune system. We will focus on Cas9 from the Type II system to understand its effector functions. We will briefly discuss the subsequent adaptations of CRISPR-Cas systems for use in scientific and clinical research. The consequences of our ability to manipulate any genome are yet to be determined. An international conversation on ethics has started but is the genie out of the bottle?

Learning Objectives

After finishing this chapter, students should be able to:

- Explain the principles of CRISPR-Cas-based adaptive immunity in bacteria and archaea.
- Describe the biochemistry of recognition and cleavage of pathogenic DNA.
- Identify some of the ethical challenges of gene manipulation.

9.1   Introduction

There are more viruses ($\sim 10^{31}$) in the world than stars ($\sim 10^{21}$–10^{24}) in the universe. Viruses outnumber bacteria by at least ten-fold. Viruses and other mobile genetic elements, such as plasmids and transposons, infect bacteria and archaea ($\sim 10^{23}$ infections per sec), per estimates @NatureRevMicro (twitter handle). To keep the infection in check, microbes have evolved various defense mechanisms including masking the receptors used by the viruses, preventing the injection of the viral DNA into the cell, or cutting the foreign DNA that enters the cells. In the CRISPR-Cas system of defense, the foreign DNA is recognized using a complementary short RNA and subsequent cutting of the pathogenic DNA. The CRISPR-Cas system is present in nearly 85% of archaea (97% in thermophiles) and 42% of the bacteria. It is an adaptive immune system where microbes build "memory" of the infection and respond to any future infections by the same pathogen using the stored information [1–6].

In the CRISPR-Cas approach to infection mitigation, microbes capture a small portion of the foreign DNA (spacer) when they first encounter a new pathogen. It is stored in an array where spacers from multiple different pathogens are stored (CRISPR array). Spacer acquisition or adaptation is the process by which Cas proteins (*CRISPR as*sociated) choose

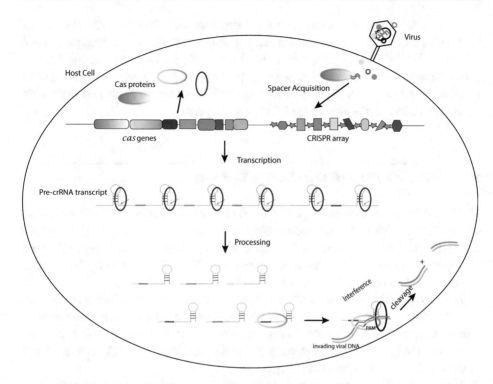

Fig. 9.1 The CRISPR-Cas-based immunity. The CRISPR-Cas system of immunity in bacteria and archaea utilizes various Cas proteins and CRISPR locus generated RNA (crRNA) to find pathogens. The proteins derived from the CRISPR-associated genes (*cas* genes) are found upstream of a locus containing short repeating sequences (teal stars). The repeats sequences are interspaced with various "spacer" sequences from different pathogens (multiple shapes and colors). When transcribed, the RNA is processed into individual crRNA. Each crRNA is loaded onto Cas proteins for surveillance of all cellular DNA. When Cas finds a particular dinucleotide PAM sequence in the DNA, it allows crRNA to bind to its complementary sequence adjacent to it. This forms a DNA:RNA hybrid strand and DNA strand that is unpaired (R-loop). The Cas protein(s) then cleave the two DNA strands of the pathogenic DNA (interference) and thus destroy the pathogen's DNA

a short segment of viral DNA to store in the CRISPR locus. The spacers are separated by repeating sequences in between them. These *c*lustered *r*egularly *i*nterspaced *s*hort *p*alindromic *r*epeats (CRISPR sequences) in the DNA are transcribed and converted into different CRISPR RNA (crRNA) (Fig. 9.1). Each crRNA is now complementary to the pathogenic DNA from which it was derived [1–4].

An operon of *cas* genes (*C*RISPR-*as*sociated) is present adjacent to the CRISPR locus to make the various Cas proteins [1–4]. The Cas proteins scan any DNA they encounter by binding to it weakly. If this Cas protein finds a specific short sequence next to a spacer that they previously acquired (the *p*rotospacer *a*djacent *m*otif—PAM), then they hydrogen bond tightly to the it. The PAM sequence wasn't stored in the host DNA and allows the Cas proteins to discriminate between the host and the pathogen's DNA.

Once the Cas proteins recognize the PAM site, they bend the DNA. This leads to the local melting of the double helical structure. It allows crRNA to start binding to DNA adjacent to the PAM site, forming a RNA:DNA helix with one strand and leaving the other strand unpaired, forming a loop-like structure (R-loop). The nucleases function of the Cas proteins cut each strand of DNA. This cleavage-based deactivation of the host DNA is the "effector function" performed by the CRISPR-Cas system.

9.2   Classification of CRISPR-Cas Systems

The CRISPR-Cas systems have been divided into two main classes. These are further divided into six different types and as many as 33 different subtypes [7, 8]. Figure 9.2 shows the *cas* gene organization in different classes and types. In class I system (Types I, III, IV), the Cas proteins are multi-protein complexes whereas class II systems (Types II, V, VI) use a single protein. Type I and III systems are abundant in archaea.

Type I CRISPR-Cas systems are the most abundant in bacteria. These systems are characterized by the formation of a large ribonucleoprotein complex, Cascade (CRISPR-associated antiviral complex for defense), that conducts RNA-guided surveillance for the presence of foreign DNA. Upon detecting and binding to the foreign DNA, it recruits Cas3, an endonuclease-helicase, to cut the target DNA.

Type II CRISPR-Cas systems have been studied more extensively as these employ a single Cas9 protein that performs both the recognition and cleavage functions. Type II CRISPR systems require a trans-acting crRNA (tracrRNA). The tracrRNA is derived from the complementary strand in a region upstream of the CRISPR locus. The tracrRNA binds to the repeat sequences near the spacer to recruit Cas9 protein and an RNase III to generate a mature crRNA. Archaea lack the RNase III nuclease required for generating the tracrRNA and therefore, lack Type II CRISPR-Cas systems.

The variety of CRISPR-Cas systems that have evolved demonstrate the co-evolution processes between the pathogen and its host. Although the details of diffrent CRISPR-Cas systems are slightly different, the key biochemical principles remain the same. The CRISPR-Cas adaptive immunity is divided into three distinct steps: the selection of the pathogenic DNA sequence that will be incorporated into the host (adaptation or spacer acquisition); generating and processing of the CRISPR RNA (expression or crRNA biogenesis); and specific binding and cleavage of the viral DNA (silencing or interference). We will discuss each of these steps below with an emphasis on Type II CRISPR-Cas9 biochemistry.

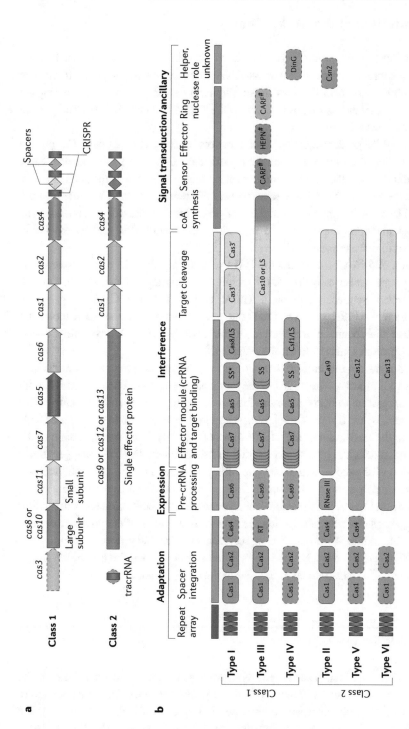

Fig. 9.2 The organization of the CRISPR-Cas systems. The modular organization for two classes of CRISPR-Cas systems seen in various organisms is shown. Class 1 have effector modules comprised of multiple Cas proteins. Class 2 have a single crRNA binding protein that is analogous to the entire set of proteins in Class 1; it requires an additional upstream RNA, the trans CRISPR RNA (tracrRNA). The tracrRNA is derived from the complementary strand. Various functional modules of the CRISPR-Cas systems and their structural, functional, and genetic organization puts them in six different types (Types I–VI). (Figure from reference [7])

9.3   Adaptation: Spacer Acquisition

The first step of developing the CRISPR-Cas adaptive immune system requires the bacteria or archaea to acquire a portion of the viral DNA from the microorganisms that infect it. Different subtypes of CRISPR-Cas differ slightly in their process of acquiring the protospacer sequences and its incorporation into the CRISPR locus [1–6, 9]. The general principles are discussed below.

The selection of the spacer sequence is not a random process. The particular Cas protein involved in the adaptation process scans for a particular ~2–5 base pair sequence, PAM (protospacer adjacent motif), on the complementary strand adjacent to the sequence that would become the spacer. Once this specific sequence is recognized the Cas protein would "acquire" the DNA on the opposite strand to convert it into a protospacer sequence. The PAM sequence plays an important role in both spacer acquisition and in the future interference as will discuss soon.

In Type-IIA CRISPR-Cas systems, the PAM-recognizing domain of Cas9 protein is responsible for protospacer selection. Cas9 recruits other proteins, Cas1, Cas 2, Csn2 to integrate the spacer sequence into the CRISPR array. Cas1 and Cas2 are generally conserved in most CRISPR systems. Cas1 and Cas2 function together as a molecular ruler and dictate the sequence architecture of the CRISPR loci. Cas1 functions as an integrase. Cas2 provides a scaffold for the complex.

A two-step integration mechanism is proposed for protospacer integration into the host DNA in the Type II systems. The 3′-OH of the protospacer end and a supercoiled DNA are necessary for integration to occur in an in-vitro assay. In the first step, Cas1 catalyzes the nucleophilic attack by the 3′-OH of the protospacer into the minus strand of the host DNA. This occurs near the leader end of the CRISPR array (Fig. 9.3). The opposite strand of the protospacer then attacks the plus strand leading to its full integration. The staggered cuts allow duplication of the repeat regions by gap filling performed by the DNA polymerase- and ligase-based repair system. Protospacer integration is observed at the borders of the each repeat adjacent to an AT-rich region. The structures of the DNA sequences in the host DNA are likely to play a role in selecting the sites of integration of the protospacer. The mechanism of integration is similar to those seen for retroviral integrases.

Some CRISPR systems use RNA and reverse transcription to acquire the spacer. In this case, the reverse transcriptase function is fused to the Cas1 protein.

9.4   CRISPR Array

After protospacer acquisition, the acquired DNA sequence is inserted into an AT-rich DNA of the host between two repeats in the CRISPR array. Once inserted, the spacer region in the array is repaired by cellular repair polymerases and ligases. CRISPR-*cas* loci contain a CRISPR array with two to several hundred direct repeats that are often palindromic. These

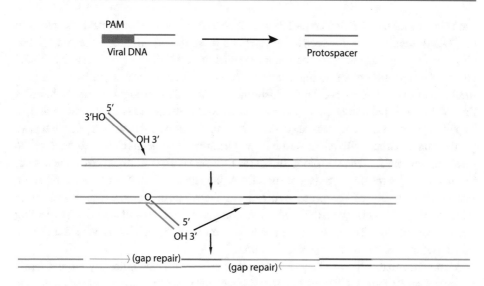

Fig. 9.3 Protospacer acquisition and integration. The protospacer is the viral DNA, adjacent to the PAM sequence. The integration of the protospacer occurs in a manner similar to the retroviral integrases. The 3′-OH acts as a nucleophile to attack the leader sequence preceding the CRISPR array. The AT-rich region bends the DNA. After integration, the repeat sequence gets duplicated on both sides of the spacer sequence by gap filling function of the host repair polymerase and ligase. In Type I systems, the process is assisted by integration host factor IHF) and in Type II systems, the leader anchoring sequence (LAS) assists in the integration process

25–35 base pairs repeats are separated by spacers. Spacers are 30–40 base pairs each. The repeats region is near the *cas* genes organized into operons (Fig. 9.2).

9.5 Creating Mature crRNA

The immune function of the CRISPR-Cas systems depends on creating a crRNA corresponding to the viral DNA that infected it [1–6, 10, 11].

The CRISPR array is first transcribed into a single, long transcript using a promoter upstream of the leader sequence. The pre-crRNA transcript is processed to produce individual crRNA, each contains one spacer sequence flanked by a partial repeat sequence. This occurs in one or two steps depending on the type of CRISPR-Cas system. An RNA hairpin structure is formed when palindromic repeat sequences are present. Mature crRNA bind to different Cas proteins in a structure-and/or sequence-dependent manner. An RNA–protein complex is formed to recognize and cleave a particular pathogenic DNA. These effector functions of the CRISPR-Cas systems vary greatly between different types.

The crRNA maturation process is very similar in Class I, Type I and Type III systems. These use a large *C*RISPR-*as*sociated *c*omplex for *a*ntiviral *d*efense (Cascade). In these systems, the pre-crRNA is cleaved within the repeat regions by Cas6 endonucleases (rarely,

Cas5) in a metal-independent and ATP-independent reaction. The Cas6 enzymes belong to the RAMP-family (repeat-associated mysterious protein) and are able to recognize and cleave a single phosphodiester bond in a repeat of the CRISPR array generating a 5′-OH and 2′-3′-cyclic phosphate termini. The reaction is likely to use a general acid-base mechanism involving an active site histidine residue. Cas6 recognizes stable hairpins formed between palindromic sequence stretches within the repeats. In *S. epidermidis*, a Type III system, crRNA are cleaved at the base of the hairpin structures within each repeat, yielding intermediate crRNA. These are further trimmed by other nucleases to produce the final mature crRNAs. The stem-loop structures themselves and the sequences at the base of the hairpin play a role in Cas6-dependent crRNA biogenesis (cartoon in Fig. 9.1 is based on this system). Many of the proteins that form the Cascade complex are involved in incorporating the correct crRNA into the complexes and simultaneously preventing crRNA degradation. The Cas6 protein remains bound to the crRNA produced and may play an additional role in the effector function.

Maturation pathways in Class 2 CRISPR systems differ significantly from Class I systems. Class 2 Type II CRISPR-Cas systems lack the Cas5 and Cas6 family of nucleases. The repeat sequences do not form a hairpin structure and therefore require a separate trans-acting RNA (tracrRNA). The sequence of tracrRNA is present on the opposite strand at a locus several nucleotides upstream of the CRISPR-Cas loci (Fig. 9.4). The tracrRNA form

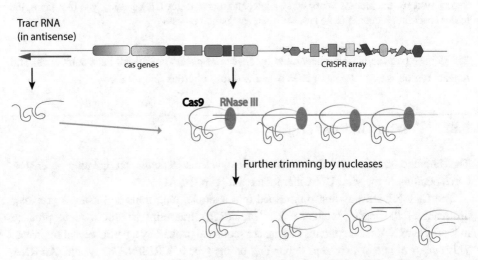

Fig. 9.4 A trans-crisper RNA (tracrRNA) binds crRNA in Class II CRISPR-Cas systems. Upstream of the CRISPR locus, a sequence on the antisense strand (green arrow) produces an RNA complementary to the repeat region in the CRISPR locus (teal stars). This RNA (tracrRNA) binds to the crRNA, with assistance from Cas9, to form tracrRNA:crRNA. This double-stranded RNA is cleaved by RNase III. A second processing step by host nucleases trims the RNA to produce a mature crRNA that exists in a ternary complex of tracrRNA:crRNA bound to Cas9 protein

a double strand RNA using the repeat sequence, a pre-crRNA:tracrRNA. The pre-crRNA:
tracrRNA is recognized and cleaved by RNase III with a two base pair separation. The
Cas9 protein promotes the duplex formation and its stabilization. Further trimming within
the spacer region occurs to form the mature crRNA. The mature crRNA consists of a
spacer-derived guide of 20 nucleotides and a repeat derived 3' half of 19–22 nucleotides.
Mature crRNA has the duplexed portion with a free 3'-OH and 5'-phosphate ends [10, 11].

The crRNA is now ready to be used as a guide RNA (gRNA) to bind its complement on
the infecting pathogen's DNA. Once the mature crRNA is produced, it binds to effector
proteins to silence the foreign DNA. The presence of different CRISPR-Cas systems and
the differences between them point to co-evolution of an organism and its pathogens.

9.6   Cleavage of Pathogenic DNA: Silencing (or Interference)

The final step of the immune response is to detect and deactivate pathogenic DNA
corresponding to the spacer sequences that are now part of the mature crRNA. The Class
2 systems utilize a single protein to perform surveillance and cleavage of the foreign DNA.
To understand the principles of this process, we will examine the Type II CRISPR-Cas
system that utilizes a single Cas9 protein for its effector function. Cas9-like proteins exist in
all Class II systems; in Class I systems, several Cas proteins are utilized to perform similar
tasks.

The effector function of the Cas9 Enzyme. The Cas9 protein from *S. pyogenes* is a large,
1368 amino-acid, multi-domain, multifunctional DNA-endonuclease. The Cas9 protein
encodes two different nucleases domains: HNH and RuvC; the HNH domain cuts the DNA
strand bound to crRNA (the target strand). The RuvC domain cuts the strand opposite to
the RNA:DNA hybrid strand that is now single-stranded (the non-target strand). The
structures of apo- and holo-Cas9 have been solved (Figs. 9.5 and 9.6) [12, 13].

The Cas9 is a bilobed enzyme that contains 25 α-helices and two β-sheets. It has a
recognition lobe (REC) and a nuclease (NUC) lobe. The REC lobe adopts a six α-helical
bundle structure with three α-helical segments (Helix I, II, and III) that are not similar to
other known proteins. The REC lobe is one of the least conserved regions of the Cas9
families within the Type II systems. The NUC lobe contains the PAM-interacting domain,
which forms an elongated seven-α-helical structure, a three-stranded antiparallel β sheet, a
five-stranded antiparallel β sheet and a two-stranded antiparallel β sheet. The fold of this
domain is also unique to Cas9 family of proteins. The NUC lobe contains the HNH domain
and a split RuvC domain along with a variable C-terminal domain (CTD). The two lobes
are connected by an arginine-rich bridge helix and a disordered linker in position 712–717
(Fig. 9.5). The CTD has a Cas9-specific fold and contains PAM-interacting sites that are
disordered in the unbound structure. Upon binding to the guide RNA, Cas9 CTD becomes
ordered and ready for recognition of PAM sites, supporting the observation that Cas9
enzymes do not function as nucleases in the absence of guide RNA.

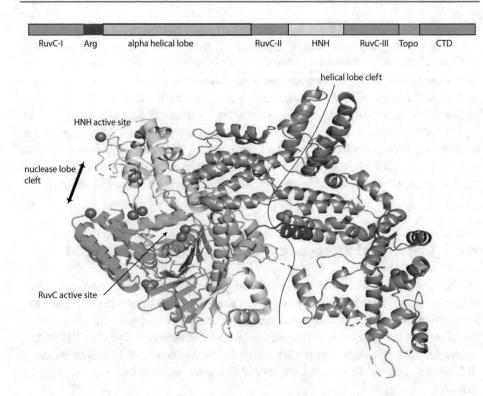

Fig. 9.5 The bilobed structure of Apo-Cas9. The domain architecture of Cas9 and the crystal structure of the apo-Cas9 protein from *S. pyogenes* are color coded to show various domains. The RuvC and HNH active sites form the nuclease domain. The two active sites are 25 Å apart. The HNH domain is poorly ordered in the structure indicating that it likely organizes upon substrate binding. A topoisomerase-like domain is found in the C-terminal with a β-β-α-β Greek key domain. The two halves of the protein are connected by the arginine-rich region (purple helix) and a linker region (residues 714–717). The figure was made using PDB file 4cmq in PyMol

The RuvC domain of Cas9 shares structural similarities with the retroviral integrase superfamily and has an RNase H fold. Therefore, it likely utilizes a two-metal mechanism for the nontarget strand cleavage. It has a conserved histidine that acts as a general base.

The HNH nuclease domain has the ββα-fold of the HNH nuclease family (even though the overall domain structure is unique to Cas9). It is likely to utilize a one-metal mechanism for the target strand cleavage and has a conserved aspartate residue (Fig. 9.7). Mutagenesis experiments of Cas9 show mutations H840A or D10A convert it into a nickase (i.e., nicks—cuts one strand) and mutating both residues simultaneously converts it to a "dead" Cas9 (dCas9), leaving the RNA-binding ability intact.

The Cas9 protein assembles with the tracrRNA:crRNA (or an artificially created single guide RNA, sgRNA, discussed in Sect. 9.8) to form a surveillance complex that can perform site-specific DNA recognition of the foreign DNA. The 20-nucleotide crRNA spacer sequence is responsible for the target recognition whereas the tracrRNA binding to

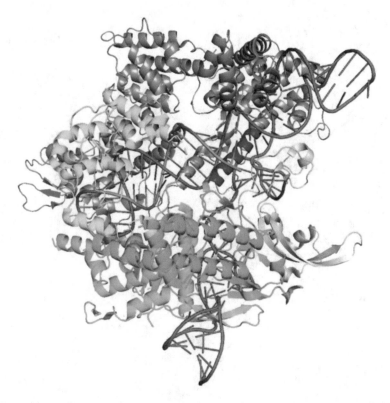

Fig. 9.6 The Holo-Cas9. The Cas9 protein bound to the guide RNA and a target DNA shows extensive interactions between the nucleic acids and the protein. The nuclease lobes (yellow and teal) contain the carboxyl-terminal domain that is involved in PAM recognition. The nuclease lobes are properly positioned for cleavage of the two DNA strands. The color coding is the same as in Fig. 9.5. The figure was made using PDB file 4OO8 in PyMol

pre-crRNA is responsible for recruiting Cas9. The PAM proximal sequences of the crRNA is on the complementary strand on the 3′-end of the 20-nucleotide space sequence. The first 8–10 nucleotide region where the crRNA binds its target (seed sequence) requires base pairing for cleavage. Any mismatches in this seed region impact the cleavage reactions by Cas9. The binding of the double-stranded RNA to Cas9 protein cause significant rearrangements in DNA recognition domain of Cas9. Helix III flips out about ~65 Å closer to the HNH domain, making room for sgRNA binding to Helix I. Much smaller changes in Cas9 are observed upon target binding, indicating that loading of the RNA is the key determinant of Cas9 activity.

Helix I, linker, and the CTD of Cas9 form extensive contacts with sgRNA, particularly in the stem-loop 1 and in the linker region between stem-loop 1 and 2. The stem-loop 2 makes some contacts with the Ruv and CTD domains. The stem-loop 3 makes very few contacts with the Cas9 protein in the holoenzyme structure. The stem-loop 1 is essential for

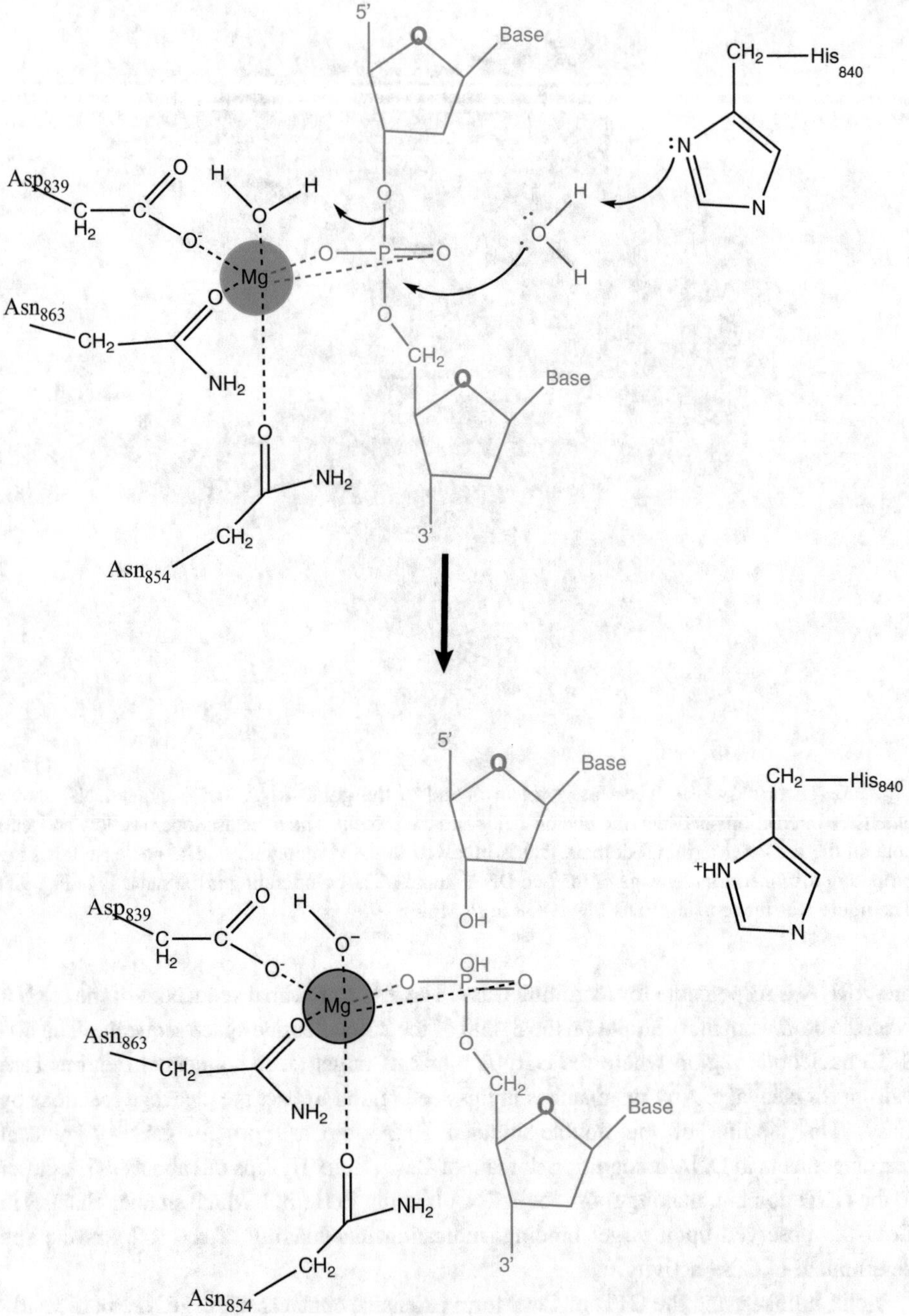

Fig. 9.7 A proposed one-metal mechanism of HNH nuclease of Cas9. The HNH nuclease residues are shown in black. The pathogenic DNA being cleaved is the target strand (red). The nucleophilic attack by water (blue) is possible upon its activation by His840 in the active site. The active site magnesium ion stabilizes the transition state

Cas9 binding and structural changes. The other regions of RNA are likely important for stabilizing the complex and for recognizing and positioning of the target DNA to form an active complex ready for optimal cleavage rates.

The Cas9 protein's interactions with the guide RNA cause preorganization of the 10-nt seed sequence for the initial DNA recognition and binding. In Type I CRISPR-Cas system, the entire guide RNA is preordered rather than just the seed region. The PAM-interacting sites on Cas9 protein, R1333 and R1335, recognize 5'-NGG-3' PAM sequence and are prepositioned to make contacts with the target DNA, allowing Cas9 to form a target-ready structure. The 5' 10-nucleotides that do not bind the target are disordered but buried in a cavity formed between HNH and RuvC domains, protecting the RNA from degradation. This implies that further changes in protein structure are needed to release 5' distal end during DNA binding.

The Cas9 bound to its double-stranded RNA is ready to search for its target, the complementary DNA that matches the seed sequence. Base pairing between the 20-nucleotide spacer sequence and the protospacer of target DNA along with a conserved PAM sequence are required. Any mutations to the PAM sequence allow the virus to escape the CRISPR-Cas-based detection. The Cas9 protein searches for the PAM sequences. Cas9 dissociates rapidly from any DNA that doesn't contain the correct PAM sequence.

The PAM containing the non-target strand interacts with the CTD domain. The first base pair in PAM doesn't interact with Cas9. The conserved GG dinucleotide binds in the major groove by base-specific hydrogen-bonding interactions with two arginine residues at 1333 and 1335 in the β-hairpin of CTD. The deoxyribose-phosphate backbone of the non-target DNA strand makes numerous hydrogen-bonding interactions with CTD.

Specific PAM binding to Cas9 triggers a sharp kink turn in the target strand immediately upstream of PAM, allowing for RNA invasion by the crRNA. This leads to the formation of the R-loop structure (Fig. 9.8) [14]. The upstream phosphodiester linkage (+1 phosphate) from the PAM site is stabilized by interactions with the K1107-S1109 residues on the protein. This phosphate-lock loop formation is crucial to destabilizing the local interactions between the two strands of DNA molecule and for flipping the first nucleotide of the target strand toward the guide RNA. This allows RNA:DNA helix to begin forming. Simultaneously, the non-target strand nucleotides are flipped out to interact with Cas9, particularly in positions −2 and −3.

Upon binding the correct PAM sequence, Cas9 causes DNA melting adjacent to the PAM site followed by the invasion of RNA strand to form an RNA:DNA hybrid. The displaced DNA strand forms a looped out structure—the R-loop (Fig. 9.8). The target DNA interacts with all 20-nucleotides of the guide RNA. A perfect complementarity leads to a predominantly A-form helix. The RNA-DNA hybrid lies in the central channel between the recognition and nuclease lobes of the Cas9 protein. It is the geometry of the helix rather than the sequence of this structure that is recognized by Cas9. The interactions of Cas9 with the single-stranded DNA induces a more pronounced conformational change in Cas9 relative to PAM binding, emphasizing the importance of RNA:DNA hybrid formation.

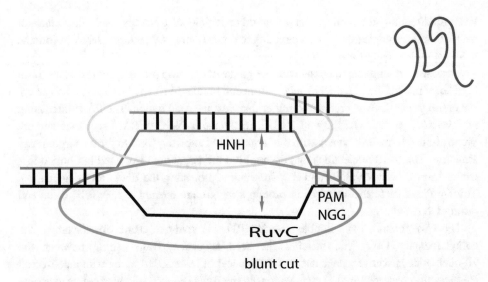

Fig. 9.8 The R-loop formation. The RNA:DNA hybrid is formed between the RNA spacer strand (teal) and the DNA strand (red). Perfect complementarity leads to an A-from helix. The RuvC domain cuts the single-stranded DNA while the HNH domain cuts the DNA in the RNA:DNA hybrid

The A-form helix of the target strand bound to crRNA threads through the central channel of the Cas9 lobes. The newly single-stranded non-target DNA strand threads into a tight side channel of the nuclease lobe. Further electrostatic interactions allow threading of the single-stranded DNA into the RuvC domain.

Overall, Cas9 bends the DNA helix and positions the DNA for crRNA binding and cleavage. In the process, Cas9 itself must undergo changes from an inactive to active state particularly in HNH domain. The allosteric changes to the loop linkers allow the non-target DNA strand to reach the RuvC active site.

The HNH domain and the RuvC domain each cleave the DNA three base pairs from the 5′-NGG PAM sequence to create a blunt end double-stranded break. The DNA binding and cleavage events of Cas9 are decoupled.

The mechanism of Cas9 illustrates that binding of mature crRNA prepares the protein for surveillance. PAM sequence recognition by Cas9 causes the R-loop formation by strand invasion. Subsequent cleavage of target and non-target strands occurs by the nuclease domains [15, 16]. Whether a single Cas9-like protein performs different functions or multiple different Cas proteins are involved differentiates the various types of CRISPR-Cas systems. Careful biochemical studies have been done on several other Cas proteins from different types of CRISPR systems.

9.7 CRISPR-Cas Evasion by Viruses

Viruses evolve with the host that they infect. It is, therefore, not surprising that viruses have evolved methods to evade the CRISPR-Cas systems. Any mutations in the PAM sequence help the virus in evading the surveillance complex. In addition, viruses have evolved to produce anti-CRISPR proteins (Acr) [17, 18]. These are 50–150 amino acids proteins that disrupt either the Cas protein's ability to bind and cleave the DNA or to interfere with the guide RNA. Over 50 different families of Acr proteins have been detected from different phages and viruses. There is little structural or sequence homology among the different Acr proteins, indicating that they have evolved independently of each other. For example, AcrIIC1 binds to the active site of the Cas proteins to prevent cleavage of the DNA whereas AcrIIC3 causes dimerization of Cas9 thus preventing its binding to the target DNA. The evolution of the host and their pathogens are tightly linked.

9.8 Gene Manipulation via CRISPR-Cas Systems

Restriction enzymes and other site-specific DNA cleavage agents (zinc-finger nucleases and transcription activator-effector nuclease) have long been of interest due to their potential to precisely cut a particular DNA whether for research purposes or for treating diseases [17–25]. The technologies developed in these systems have been used to transform molecular biology and have shown great potential for gene editing of organisms.

The genius of the Charpentier and Doudna laboratories was to artificially fuse the tracrRNA to any spacer sequence thus, providing a single guide RNA (sgRNA) that could be tailored to bind any target DNA (Fig. 9.9). The ability of CRISPR-Cas9 system to target any potential DNA sequence opened up the possibilities for genome modifications. CRISPR-Cas9 stands out from all the prior attempts to perform site-specific target recognition and high efficacy DNA cleavage by its simple design, ease of use, and cost effectiveness.

The heart of CRISPR recognition lies in a complementary RNA sequence that leads a Cas9(-like) protein to a specific locus on the genome for cleavage. In addition, cleavage non-competent Cas9 (dead Cas9, dCas9) can be used for sequence-specific DNA binding for cell imaging, transcriptional control, or many of the other applications that requires site-specific tagging of a DNA site with a protein.

When cellular systems are damaged, there are (at least) two ways that the cells repair the damaged genes. In dividing cells, homologous recombination utilizes the second chromosome to repair the damaged chromosome [26]. In non-dividing cells, non-homologous end joining utilizes repair enzymes to insert in arbitrary nucleotides to join the broken double helical ends [27]. The cells' inherent repair mechanisms can be utilized to repair the DNA cleaved by CRISPR-Cas system. A new or modified gene sequence can be provided for repairing the DNA using homologous recombination. This allows gene manipulation in a manner previously only imagined in science fiction. Now with CRISPR-Cas, genes can be

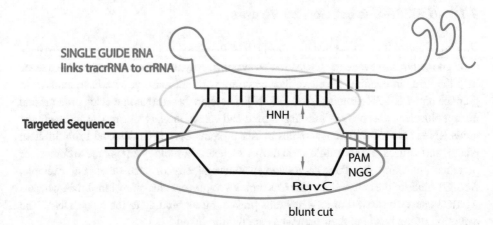

SINGLE GUIDE RNA
links tracrRNA to crRNA

Targeted Sequence

HNH

PAM
NGG

RuvC

blunt cut

Fig. 9.9 Targeting DNA using sgRNA. The tracrRNA is linked to a sequence complementary to any target gene (magenta). This single guide RNA (sgRNA) is loaded on to Cas9 to target the gene of interest (dark blue). The sequence of sgRNA can be designed with just 20–30 nucleotide change in the sgRNA. This allows one to target any sequence of interest, thus making it an ideal tool for genetic manipulation

cut and removed *or altered*. Now genes can be copied and pasted almost as easily and cheaply as editing a textbook, opening up endless possibilities of "messing" with the genome. We now have the technology to change the makeup of any organism and to cure diseases that were proving difficult. This is the new era of personalized medicine with the potential to literally change us.

9.9   Ethics of Gene Manipulation

Having the power to alter any organism, including ourselves, raises profound questions [28–40]. The stakes in our understanding of biology and in our experimental procedures could not be any higher. Our limited understanding of the complexity of biological processes and interconnections between organisms can have a profound effect on life on earth. The potential for benefit and harm from this technology and layers of ramifications are unfathomable! Dr. Doudna has raised some of these concerns [28]. The ethical and regulatory bodies have begun the conversations to build some international policies around CRISPR-Cas-based genome manipulations [28, 29, 35, 38, 39].

Our understanding of the basic principles of biology paved the way for CRISPR-Cas system to become an indispensable tool of molecular biology. Use of protein- and nucleic acid-enzymes for site-specific cleavage had been explored for decades before the discovery of CRIRPR-Cas. The expense of designing each site-specific cleavage has kept these technologies from being easy or cheap to use, thus limiting their widespread use. Several large developments have changed the landscape of biological research in the meantime.

The human genome sequencing made genomic information on humans and other organisms readily available. The DNA itself became cheap. Targeting of genes is now possible with greater precision, even while using older technologies. Altering of gene sequences to study the role of any gene for research experiments is more straightforward. The availability of CRISPR-Cas as tool for gene editing in the current times allowed an explosion in CRISPR-associated applications for biological research, medical research and more. Not surprisingly, it led to Nobel prizes for Dr. Jennifer Doudna and Dr. Emmanuelle Charpentier. Many new technologies based on CRISPR-Cas systems are being developed, some using dCas9, that allow, targeting sequences without cleavage. Many scientists are thinking about the ethical implications of genome editing to determine new ways to use the technology that will not cause harm to humans or other organisms.

Since its discovery, CRISPR-Cas-based technology has been used to manipulate large and small organisms. For example, CRISPR-Cas technologies have been used to enhance muscles in dogs, alter the size of pigs, edit the genes in rice, corn, soybean, tomatoes among many other organisms and has raised concerns internationally. Arguably, the technologies for genetic alteration of animals and plants have been with us for a long time, with animal husbandry and grafting in plant. *The scale and speed at which gene alteration is now possible with CRISPR-Cas systems is unimaginable* (and uncontrollable.)

Below are some issues that I have compiled for these discussions. The issues raised here are not meant to be comprehensive. This is a starting point to have further conversations on genome alterations (using any technology). We all need to participate in conversations around human genetic modification and policy developments. We need to ask how and when questions along with the trade-offs that we may find acceptable.

Complexity of biological systems. Our understanding of biology/biochemistry of an organism and interconnection between species is vast and yet very limited. A key concern in any in-vivo genetic manipulation is the *accuracy and precision* of the method. To precisely alter what we want to, we need to ask questions like: Does the guide RNA (gRNA) sequence bind to a single site on the genome as designed or are there unintended targets (i.e., non-target specificity)? Can a seed sequence bind to a site in a novel, non-complementary manner to make alternate structures that are viable for cleavage? What sub-optimal structures might form in-vivo that are sufficient for cleavage by Cas protein or other nucleases? How often does this occur? Undoubtedly, some of the questions are being answered by current studies. As these technologies are being patented, information is moving away from the public domain of academic research, which raises systemic problems around access, informed consent, discrimination etc. *In addition, there are many questions that we do not yet know to ask.*

Technological advances in protein engineering are making high-fidelity enzymes possible—enzymes that are more precise. However, non-specific target cleavage in large genomes is a non-trivial problem. New methods of anti-CRISPR proteins (Acr) are being used to fine tune the cleavage reaction and to "destroy" the CRISPR-Cas system after a set amount of time—thus, decreasing unintended, slow reactions to go on. These methods show promise as non-specific cleavage is decreased. Even as some technological challenges are overcome, questions will remain on other fronts.

Oncogenes or other potentially slow(er) effects. Could CRISPR targeting directly or indirectly activate or inactivate genes that control cells from becoming cancerous—turning on or off oncogenes? Do we know the full landscape of oncogenes, along with their multiple and complex interconnections with other pathways? If CRISPR-Cas systems destroys a gene, like p53, that prevents cancer or creates an oncogene, then the time frame for the effects of such modifications is unclear. The long-term impacts of CRISPR-Cas will be hard to monitor. Any correlation between diseases and its causes will require closely monitoring those receiving treatment (along with a control cohort) over a long period of time.

Single gene, multiple functions. The potential that defective genes could be cut and removed, opens up exciting possibilities for treating diseases, especially those being caused by a single or very few defective genes. We have the power to ease human suffering from previously untreatable diseases. We know that often a single gene does not produce a single protein, and a single protein does not have a single function, as we once thought. Now we know a single sequence of DNA produces many mRNA isoforms, which produce many proteins with functions that are cell and tissue specific (Chap. 4, Spliceosome). When the DNA is modified for one function, it will unintentionally alter other functions connected to it. The gene for protein PCSK9 (proprotein convertase subtilisin/kexin type 9) produces a protein that is defective in the case of familial hypercholesterolemia. PCSK9 binds to receptors responsible for breaking down artery clogging low-density lipoproteins (LDL). Without PCSK9, the cells have more LDL to remove cholesterol from the blood. This protein was recently targeted successfully in monkeys using CRISPR-Cas system; it showed 60% decrease in "bad" cholesterol for at least 8 months. However, PCSK9 has other roles in immunity; it is also expressed in testis and pancreas. What effect does CRISPR-Cas modifications have on these other functions of PSK9 and associated systems? Is this gene responsible for expression of other genes? Would removal or alteration of gene have other systemic effects? If the unintended consequences are acceptable in the light of severity of the disease, then should we permanently eliminate the defect by editing gametes (egg or sperm cells)?

The complexity of these issues is magnified when an individuals can be born with mosaic genetic combinations (modified and unmodified cells in the same organism) as is the case with the ΔCCR5 modification first reported in the case of CRISPR-Cas modified babies born in China in 2017.

The Public Landscape of Decision Making. Some of the questions that the CRISPR-Cas technique raises have been discussed for decades, since the early days of gene transfer technology. Other questions arose with the discovery and rise of stem-cell-based therapies. A key element that has been missing in use of these technologies has been any significant public input. The current environment in which masks and vaccines are controversial have highlighted our illogical nature and the role of politics in science. The public opinion surveys indicate current support for modifications that are temporary (somatic cells; one generation) versus those that are permanent (germ cells; heritable).

Often, people are assumed to be making these decisions for themselves or their children. It is unclear what role the insurance companies or employers will play in these areas.

Recent fear that the current coronavirus pandemic was caused by a laboratory created modified organism has shown a level of distrust in science and scientists that may have only been seen in movies before. The CRISPR-Cas system could potentially contribute to a similar scare were it to become linked to the technology rather than the complex issues that surround the use of any technology. The controversies around stem-cell research are worth revisiting.

Gain-in-function modifications. Should we be pursuing research in areas where CRISPR-Cas can prevent diseases (before they manifest) or should it be used after the disease and its impact on a person or the society are known? This leads us to the ideas of gain-in-function modifications to prevent diseases. Are modifications just for the prevention of diseases or should athletes be allowed to gain muscle? Should we use the technology to improve brain functions generally or to treat ADHD or autism? One could argue that a healthy body and a healthy brain will make better decisions. When will we cross a line into eugenics?

Genome Editing (Non-Human) Organisms The initial plasmids-based gene manipulation started with modifications in bacteriophage, bacteria, and other microorganisms. An ethical and regulatory framework for such genetically modified organisms was developed with specific criteria for use, storage, and disposal of cloned cells. The practice is now so routine that we do not stop to think about the manipulation of life that we are currently undertaking. These molecular methodologies have transformed biological sciences and resulted in numerous life-saving technologies, for example, we produce human insulin for the treatment of diabetes inside bacterial cells via transgenic techniques.

Capitalism and Access to Information. Certain controversial applications of genetic manipulation are under the radar but may provide insights into the new era of CRISPR-Cas-based editing of organisms. For example, large corporations have patented seeds solely for monetary gains; it has caused significant harm to the farming communities with little evidence of increase in production over time. The conversation on genetically modified (GM) food and its impact on plant and microbial communities, soil, pesticide use, and the environment are difficult to monitor as these are complex systems. Corporations are founded on principles of economic gain therefore, they disenfranchise groups that do not serve their economic interests; they have a strong arm for lobbying governments. All these tactics keep implications of these technologies out of the public's reach. The safety of eating GM foods is a popular topic for public discourse but perhaps, scientifically, it is the least controversial dimension of the technology. (Dr. Doudna and others in CRISPR field have already encountered those with economic interests in commercializing designer babies.)

Experiments have been performed on mosquitoes to control their fertility using genetic manipulations (gene-drives) to prevent human diseases with limited success. Ecological studies on large predators also provide a cautionary tale on manipulating the balance of a natural system.

Human diversity and equity, informed consent, and definitions of disability. The use of cancerous cells from Henrietta Lacks without her (or her families) consent (or associated financial benefits) brought issues of diversity, equity and informed consent to the forefront. Much medical literature exists around informed consent and its limitations.

When discussing diseases, the definition of disability is paramount. Those who have extra chromosomes (Down's syndrome), hearing- or sight-related challenges contribute to improvement of society in many different ways. Diverse perspectives add to our collective humanity and our understanding of the world. If we have the potential to "cure" these diseases or alter humans based on a narrow definition of humanity than is the world better or worse for it? Would having a more homogeneous population lead to more devastating future for infectious diseases? Who gets to decide this for whom? For example, our understanding of sex as exclusively male or female has caused significant harm (via medical practices) and our understanding of sexual diversity.

Access. As with any conversation about technologies, issues of access remain paramount. Current medical interventions are not available equitably to all those who need it. When treatments that are created by taxpayer funded programs are available to a few individuals—those with extraordinary health insurance or those who are independently wealthy—it makes these technologies unavailable to a large majority of people, exacerbating the differences between haves and have-nots. Sex-determination, in vitro fertilization (IVF) technologies, expensive antibody treatments, and cancer therapies have all demonstrated the role access and wealth play in a society. What procedure (and for whom) would insurance cover different treatments is one of the many conversations that scientists need to participate in.

The role of education. If the concerns regarding the coronavirus vaccine and associated myths are anything to go by, then the use of CRISPR-based technologies will need both better science and a much better education for the public. Are we as scientists prepared for the challenges that lie ahead?

Life on earth has evolved over billions of years and an ecological balance is something that cannot be easily predicted or controlled. When a technology has potential to change organisms, bring back extinct animals (example, creating mammoths from elephants), make designer babies along with providing numerous benefits, including treating diseases, providing sufficient food for the planet, protecting the climate, then how do different societies make their decisions?

Take Home Message
- Different CRISPR-Cas systems utilize similar principles for adaptation and effector functions but may utilize different Cas proteins for the tasks.

(continued)

- The biochemical principles of DNA cleavage, strand invasion, DNA repair are similar in different organisms, showing the evolutionary conservation of key mechanisms.
- Ethical challenges posed by CRISPR-Cas-based genetic modifications are formidable and will be challenging to navigate. It is paramount that we all engage in the conversations around ethics of gene editing before it is too late.

References

1. Sorek R, Lawrence CM, Wiedenheft B. CRISPR-mediated adaptive immune systems in bacteria and archaea. Annu Rev Biochem. 2013;82:237–66.
2. Jiang F, Doudna JA. CRISPR-Cas9 structures and mechanisms. Annu Rev Biophys. 2017;46: 505–29.
3. Doudna JA, Charpentier E. The new frontier of genome engineering with CRISPR-Cas9. Science. 2014;346:1077.
4. Hille F, Charpentier E. CRISPR-Cas: biology, mechanisms and relevance. Philos Trans R Soc B. 2018;371:20150496.
5. Nuñez J, Lee A, Engelman A, et al. Integrase-mediated spacer acquisition during CRISPR–Cas adaptive immunity. Nature. 2015;519:193–8.
6. Barrangou E, van der Oost J, editors. CRISPR-Cas systems: RNA mediated immunity in bacteria and archaea. New York: Springer; 2013.
7. Makarova KS, Wolf YI, Iranzo J, et al. Evolutionary classification of CRISPR–Cas systems: a burst of class 2 and derived variants. Nat Rev Microbiol. 2020;18:67–83.
8. Koonin EV, Makarova KS. Origins and evolution of CRISPR-Cas systems. Philos Trans R Soc B. 2019;374:20180087.
9. Nuñez JK, Kranzusch PJ, Noeske J, et al. Cas1–Cas2 complex formation mediates spacer acquisition during CRISPR–Cas adaptive immunity. Nat Struct Mol Biol. 2014;21:528–34.
10. Deltcheva E, Chylinski K, Sharma CM, et al. CRISPR RNA maturation by trans-encoded small RNA and host factor RNase III. Nature. 2011;471:602–7.
11. Charpentier E, van der Oost J, White MF. crRNA biogenesis. In: Barrangou E, van der Oost J, editors. CRISPR-Cas systems, chap 5. New York: Springer; 2013. p. 115–45.
12. Jinek M, Jiang F, Taylor DW, et al. Structures of Cas9 endonucleases reveal RNA-mediated conformational activation. Science. 2014;343:47997.
13. Nishimasu H, Ann Ran F, Hsu PD, et al. Crystal structure of Cas9 in complex with guide RNA and target DNA. Cell. 2014;156:935–49.
14. Jiang F, Taylor DW, Chen JS, et al. Structures of a CRISPR-Cas9 R-loop complex primed for DNA cleavage. Science. 2016;351:867–71.
15. Redding S, Sternberg SH, Marshall M, et al. Surveillance and processing of foreign DNA by the E. coli CRISPR-Cas system. Cell. 2015;163:854–65.
16. Gasinunas G, Barrangou R, Horvarth P, et al. Cas9-crRNA ribonucleoprotein complex mediates specific DNA cleavage for adaptive immunity in bacteria. Proc Natl Acad Sci. 2012;109: E2579–86.
17. Peng X, Mayo-Muñoz D, Bhoombalan-Chitty Y, et al. Anti-CRISPR proteins in archaea. Trends Microbiol. 2020;28:913–21.

18. Liu Q, Zhang J, Huang X. Anti-CRISPR proteins targeting the CRISPR-Cas system enrich the toolkit for genetic engineering. FEBS J. 2020;287:626–44.
19. Venken KJT, Bellen HJ. Emerging technologies for gene manipulation in Drosophila melanogaster. Nat Rev Genet. 2005;6:167–78.
20. Gaj T, Gersbach C, Barbas CF III. ZFN, TALEN and CRISPR/Cas-based methods for genome engineering. Trends Biotechnol. 2013;31:397–405.
21. Wright AV, Nunez JK, Doudna JA. Biology and applications of CRISPR systems: harnessing nature's toolbox for genome engineering. Cell. 2016;164:29–44.
22. Komor AC, Badran AH, Liu DR. CRISPR-based technologies for the manipulation of eularyotic genomes. Cell. 2017;169:559.
23. Hsu PD, Lander ES, Zhang F. Development and applications of CRISPR-Cas9 for genome engineering. Cell. 2014;157:1262–78.
24. Cameron P, Coons MM, Klompe SE, et al. Harnessing type I CRISPR-Cas systems for genome engineering in human cells. Nat Biotechnol. 2019;37:1471–7.
25. Murugan K, Babu K, Sundaresan R, et al. The revolution continues: newly discovered systems expand the CRISPR-Cas toolkit. Mol Cell. 2017;68:15–25.
26. Li X, Heyer WD. Homologous recombination in DNA repair and DNA damage tolerance. Cell Res. 2008;18:99–113.
27. Chang HHY, Pannunzio NR, Adachi N, et al. Non-homolgous DNA end joining and alternative pathways to double-strand break repair. Nat Rev Mol Cell Biol. 2017;18:495–506.
28. Sternberg S, Doudna JA. A crack in the creating: gene editing and the unthinkable power to control evolution. Boston: Houghton Mifflin Harcourt; 2018.
29. National Academy of Medicine, National Academy of Sciences, and the Royal Society. Heritable human genome editing. Washington, DC: The National Academies Press; 2020.
30. Baylis F, Alterted Inheritance. CRISPR and ethics of human genome editing. Harvard University Press, Boston; 2019.
31. Editorial. A CRISPR definition of genetic modification. Nat Plants. 2018;4:233.
32. Reardon S. CRISPR gene-editing creates wave of exotic model organisms. Nature. 2019;568: 441–2.
33. McConnell SC, Blasimme A. Ethics, values and responsibility in human genome editing. AMA J Ethics. 2019;21:E1017–20.
34. Sugarman J. Ethics and germline gene editing. EMBO Rep. 2015;16:879–80.
35. International Bioethics Committee. Report of the International Bioethics Committee (IBC) on updating its reflection on the human genome and human rights. Final recommendations. Rev Derecho Genoma Hum. 2015;43:195–9.
36. Pollack R. Eugenics lurk in the shadow of CRISPR. Science. 2015;348:871.
37. Lander ES, Baylis F, Zhang F, et al. Adopt a moratorium on heritable genome editing. Nature. 2019;567:165–8.
38. Wolinetz C, Collins F. NIH supports call for moratorium on clinical uses of germline gene editing. Nature. 2019;567:175.
39. Dzau VJ, McNutt M, Ramakrishnan V. Academies' action plan for germline editing. Nat Cell Biol. 2019;21:1479–89.
40. Jasanoff S, Hurlbut JB, Saha K. CRISPR democracy: gene editing and the need for inclusive deliberation. Issues Sci Technol. 2015;32:37.

10

Rachel Wilson, Evgenia Shishkova, Chris Dickinson, Jordyn M. Wilcox, Natalie L. Nicholls, Andy J. Wowor, Hayden Low, Neena Grover, and Jennifer F. Garcia

Contents

R. Wilson · E. Shishkova · C. Dickinson · N. L. Nicholls · A. J. Wowor · N. Grover
Department of Chemistry and Biochemistry, Colorado College, Colorado Springs, CO, USA
e-mail: r_wilson@coloradocollege.edu; Christopher.Dickinson@coloradocollege.edu;
ngrover@coloradocollege.edu

J. M. Wilcox · H. Low · J. F. Garcia (✉)
Department of Molecular and Cellular Biology, Colorado College, Colorado Springs, CO, USA
e-mail: h_low@coloradocollege.edu; jgarcia@ColoradoCollege.edu

© Springer Nature Switzerland AG 2022
N. Grover (ed.), *Fundamentals of RNA Structure and Function*, Learning Materials in
Biosciences, https://doi.org/10.1007/978-3-030-90214-8_10

Keywords

RNA initiation · RNA elongation · RNA termination · RNAP · RNA Pol II

What You Will Learn

Transcription plays an essential role in producing protein-coding messenger RNAs (mRNAs) and also other noncoding RNAs essential for proper cell function such as tRNAs, rRNAs, and miRNAs. A large protein complex, RNA polymerase (RNAP), catalyzes the necessary enzymatic reactions to create polymers of ribonucleotides in a manner that is complementary to the DNA templates they are transcribed from.

First, you will learn about the enzyme RNA polymerase (RNAP) and how it is a key enzyme found in all forms of life. RNAP is responsible for mediating the production of RNA transcripts from a DNA template in a process called transcription. Remarkably, RNAPs from prokaryotes, archaea, and eukaryotes are highly conserved in structure, function and enzymatic activity. We will explore the enzymatic reaction catalyzed by RNAP. Next, we will become familiar with the three different phases of transcription: initiation, elongation, and termination. Lastly, you will learn about how bacterial RNAPs and the eukaryotic RNAP II recognizes where it can initiate RNA synthesis. You will learn that transcription initiation requires the recognition of various DNA elements found within the promoter of a gene. These proteins, such as the bacterial sigma factor or the RNAPII general transcription factors, associate with the promoter to load and prime RNAP for transcription.

Learning Objectives

After finishing this chapter, should be able to:

- Describe the conserved structure and function of the core enzyme that mediates transcription, RNA Polymerase.
- Compare and contrast transcription by bacterial RNAP and RNAPII. Describe steps and factors necessary for transcription initiation, elongation and termination.
- Describe how promoters are recognized by bacterial RNAP and RNAPII.

10.1 The Structure and Enzymatic Activity of RNAP

10.1.1 The Basic Structure of the Core Enzyme

The RNA polymerases (RNAP) are large DNA-dependent protein assemblies that are responsible for the transcription of RNA in prokaryotes, archaea, and eukaryotes. Varying forms of RNA polymerase exists in each type of cell. For example, in most bacteria, RNAP is made up of five different subunits while the archaeal RNAP is composed of 11 subunits. While, the RNAPs found in eukaryotes have 12–17 subunits. Although each RNAP differs in subunit complexity, all RNAPs have been found to have the same essential core structure where all RNAPs share a common core set of five highly related subunits.

To describe the basic structure of the core enzyme for RNA synthesis, we will focus on the structure of the bacterial RNAP core enzyme and the eukaryotic RNA polymerase II (RNAP II). Both structures of the bacterial RNAP [1] and RNAP II core enzyme have been solved [2, 3] and can be used as a model for the core enzyme of other RNAPs. Bacteria have only one RNAP, which can transcribe all classes of RNA. Bacterial RNAP, with a mass of ~400 kDa, consists of five subunits (α_1, α_2, β, β', and ω; Fig. 10.1). RNAPII, has a mass of 514 kDa and consists of 12 subunits. Of those 12, 5 make up the core enzyme (Rbp1, Rbp2, Rbp3, Rbp11, and Rbp6). Rbp1 is homologous to β' while Rbp2 is homologous β; and together make up the catalytic core. Rbp3 and Rbp11 are homologous to the two α subunits, which form a homodimer that is involved in RNAP assembly and helps mediate DNA–RNAP interactions. Lastly, Rbp6 is homologous to ω and is important for RNAP enzyme folding (Fig. 10.1). Both the bacterial RNAP and RNAPII fold into a claw-like structure where one half of the claw is formed mainly by the β/Rbp2 subunit. The other half of the claw is formed mainly by the β'/Rbp1 subunits. The two α-like subunits are found at the hinge of the claw. The ω subunit is found to directly interact with β' and located on the opposite edge of where β' interacts with β (Fig. 10.1).

An active site is contained within the cleft formed by β and β'-like subunits of RNAP. The β'-like peptide sequence contains three aspartate residues, which coordinate a Mg^{2+} ion (termed MgA), and form the catalytic core that mediates the NTP addition reaction to make the RNA transcript. Lastly, the catalytic core requires a second Mg^{2+} ion, called MgB, that is brought to the active site by the incoming NTP and coordinated by the residue Asp837 of Rpb2 [4].

Three distinct openings in RNAP allow for access to the active site and are apparent in crystal structures of both the eukaryotic and bacterial RNAPs. The main channel is a positively charged groove that binds the downstream DNA along the cleft of the claw and allows the DNA to enter into the active site. An RNA exit channel is present and branches off the DNA binding groove close to the active site. Lastly, a secondary pore is postulated to allow for the entry of NTPs to the active site [5].

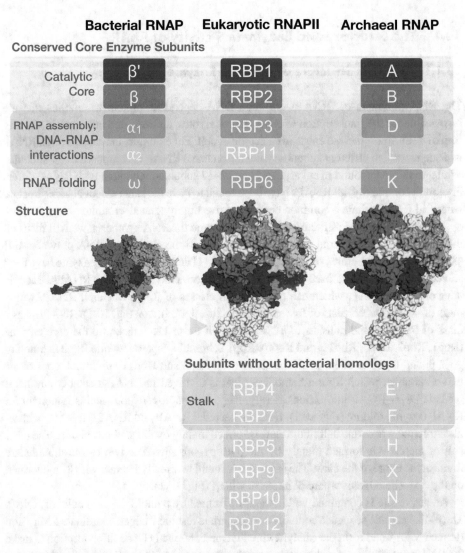

Fig. 10.1 Top. Table denoting conserved core RNAP subunits and function. Bottom. Table denoting auxiliary subunits of eukaryotic RNAP II and archaeal RNAP

10.1.2 Major Differences in Basic Structure of the Core Enzyme

Though studies have revealed high levels of RNAP structural conservation across all domains of life, many of the RNAPs slightly differ from each other. For example, Archaea, like prokaryotes, contain only one RNAP. However, unlike the bacterial RNAP, the archaeal RNAP is composed of 11 subunits. The α_1, α_2, β, β', and ω-like subunits (denoted A, B, D, L and K subunits, respectively; Fig. 10.1) are homologous to the five subunits found in bacteria. The most notable difference between archaeal RNAP and bacterial RNAP is the appearance of a stalk-like structure formed by subunits E and F in the archaeal RNAP (green arrow, Fig. 10.1). The stalk in archaeal RNAP seems to act as a scaffold capable of stabilizing the nascent RNA as the E and F subunit complex binds to RNAP near the RNA exit channel [6]. Secondly, the E and F subunits are proposed to strengthen the ability of RNAP to bind more tightly to the DNA during transcription as these subunits preferentially associate with the closed RNAP, a form of RNAP that is known to bind tighter to DNA after promoter recognition [6]. Archaeal RNAP is often described as a simplified version of eukaryotic RNA polymerase II because RNAPII also contains a stalk and many of the additional subunits found in the Archaeal RNAP are highly homologous to eukaryotic RNAPs (Fig. 10.1 Green arrow). For example, the archaeal RNAP subunit E of the stalk is highly conserved to the RNAPII subunit, Rbp7.

10.1.3 The General Reaction Mechanism of RNAPs

The primary activity of RNAP is to catalyze phosphodiester bond formation between an RNA polymer and ribonucleotides in a DNA templated manner to generate protein-coding mRNAs or non-protein-coding RNAs. However, for RNAP to perform this reaction, multiple steps must be achieved prior to and after phosphodiester bond formation to successfully produce an RNA transcript competent for translation or noncoding RNA function.

The multistep process of RNA synthesis can be broken down into three distinct phases: transcription initiation, elongation, and termination. During transcription initiation, by and large, RNAPs must recognize and bind to specific DNA elements that mark the start of transcription, called promoters. Then, productive RNAP binding to the promoter induces melting of the double stranded DNA at the promoter. This allows for the active site within RNAP to access the DNA template. Lastly, RNAP makes another transition during transcription initiation to break the DNA-protein contacts made to recognize the promoter to initiate RNA synthesis and travel along DNA, a critical step called promoter escape. RNAP then transitions into the transcription elongation phase where RNAP catalyzes a processive reaction to add NTPs to the $3'$ end of RNA polymers. To do this, RNAP reads the DNA template (traveling along the template strand in a $3'$ to $5'$ direction) and engages the correct complementary NTP within the active site. Once the correct NTP is within the catalytic core, RNAP catalyzes phosphodiester bond formation between $5'$ α-phosphate of the incoming NTP and terminal $3'$ hydroxyl of the growing RNA polymer. This reaction is energetically favorable and releases a pyrophosphate. Coupled with phosphodiester bond formation and pyrophosphate release is the translocation of RNAP along the DNA template to the next base in the DNA template, where the highly processive RNAP can repeat the reaction. During transcription elongation, RNA synthesis is signaled to stop, inducing the last phase of transcription called transcription termination. Here, RNAP disassembles from the DNA and releases the nascent RNA transcript.

10.1.4 Nucleotide Addition Cycle

To transcribe the RNA transcript properly, RNAP must take great care in selecting which nucleotides to incorporate into the RNA molecule. Errors in this process can lead to RNA molecules that encode improper amino acids for protein synthesis or RNAs with different secondary structures; both outcomes can compromise the function of the gene product. To facilitate proper nucleotide selection, RNAPs unwind the double stranded DNA strand near the transcriptional start site (e.g., the TSS or the +1 site; the site where RNA synthesis starts). The two unwound strands of DNA create a transcriptional bubble composed of a single-stranded template strand and the single-stranded coding strand where approximately 12–14 base pairs of DNA is melted. Once the transcriptional bubble is formed, RNAP has the DNA template strand within its active site that it can use to template NTP addition.

The active site contains two critical structural elements: the trigger loop, and the bridge helix (Fig. 10.2). The trigger loop (residues 1070–1100 of Rpb1) is a mobile helix-loop-helix element important for sensing the correct NTP in the active site. The bridge helix is an α-helix (residues 815–845 of Rpb1) that spans the cleft between Rpb1 and Rpb2 and senses conformational changes within the trigger loop to mediate proper nucleotide selection, catalysis and translocation. Both elements are highly conserved in all RNAPs and play major roles in adding NTPs to the nascent RNA molecule in a process called the Nucleotide Addition Cycle [7].

Catalysis by the active site requires two catalytic Mg^{2+} cations, called MgA which is coordinated in the active site; and MgB, which is brought in by the incoming NTP (Fig. 10.2). Many contacts between the correct NTP and RNAP are formed and used to discriminate that the proper nucleotide has been added. First, Watson-Crick base pairing between the NTP and the DNA template form hydrogen bonds. Second, to discriminate between ribonucleotides (rNTPs) and deoxyribonucleotides, Asn479 and Gln1078 of Rpb1 closely interact with NTPs lacking a 2′ OH group [4]. Arg766 and Arg1020 in Rpb2 hydrogen bond with the γ-phosphate of NTP to help orient the NTP for catalysis [8]. Lastly, His1085 hydrogen bonds with the β-phosphate of the NTP and is thought to help position the β-phosphate near the 3′ OH of the nascent RNA for catalysis [4].

The rNTP to be added plus the MgB ion can only access the active site when the trigger loop is in an open conformation (Fig. 10.2). Upon binding of the correct NTP to the active site, the trigger loop undergoes a conformational change to the "closed" loop conformation. The closed loop conformation of the trigger loop causes the trigger loop to sterically push against the bridge helix, causing the straight alpha helix of the bridge helix to "kink." Also, the closed trigger loop facilitates a favorable intermediate where the α-phosphate of the NTP is brought in close proximity to the 3′ OH of the growing RNA end, thus, potentially catalyzing phosphodiester bond formation to add the rNTP on the 3′ end and the release of a pyrophosphate. Simultaneously, the kinked bridge helix also induces a conformational change in RNAP that causes the RNAP to translocate to the next base in the DNA template, moving on the DNA template strand 3′ to 5′. Once the NTP has been added and the RNAP translocates to the next base of the DNA template into the active site, the trigger loop then

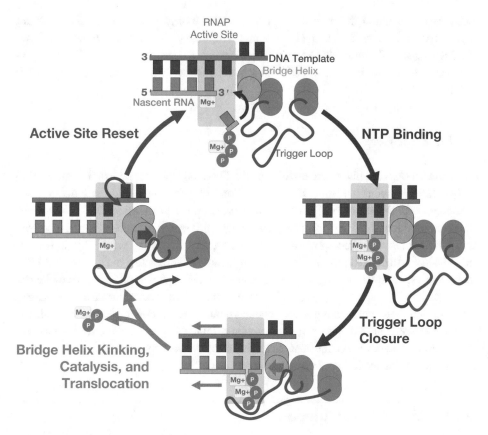

Fig. 10.2 A diagram depicting the nucleotide addition cycle

adopts the "open" loop conformation and the bridge helix relaxes back into a straight helical conformation [9]. This restarts the nucleotide addition cycle, where the active site is open for another NTP (Fig. 10.2).

10.1.5 The Elongating RNAP Is a Processive Enzyme

RNAP facilitates RNA synthesis in a unidirectional manner traveling on the DNA template strand 3′ to 5′. The elongating RNAP is able to consecutively execute nucleotide addition over and over again without releasing the nascent RNA molecule or the DNA template. Because of this ability of RNA polymerase to execute repeated nucleotide additions without releasing its substrate, an elongating RNAP can be considered a processive enzyme. This enables RNAP to transcribe long transcripts without letting go of the nascent RNA prior to the end of a gene. One key and highly conserved structural feature of RNAPs that promotes processivity is called the Rudder. The Rudder is a part of the β′ and Rbp1 subunits and contacts the nascent RNA to stabilize its association with RNAP. Mutant RNAPs without the rudder element, are more susceptible to disassociate from the RNA and the DNA than wild-type RNAPs [10]. Additionally, the processivity of RNAPs can be enhanced by factors that associated with RNAP, such as NusG in bacteria and Spt5 in eukaryotes.

10.1.6 Error Rate

Although RNAPs are highly processive, the fidelity of all RNAP is far lower than the DNA polymerases responsible for DNA replication. On average RNA polymerase will misincorporate the wrong nucleotide at a rate 10,000 times greater than DNA polymerase. It is estimated for RNAP to misincorporate 1 nucleotide every 10^6 nucleotides. In comparison DNA polymerase incorporates one mistake every 10^{10} base pairs. This is because RNAPs lack a dedicated proofreading domain capable of exonucleolytic activity like DNA polymerases. Despite lacking a proofreading domain, a structural study from Patrick Cramer's lab has shown that RNAPs are capable of correcting mistakes by pausing transcription then backtracking to the misincorporated nucleotide to remove of the incorrect nucleotide [11]. This pause in transcription is induced by misincorporated nucleotide, which slows the addition of the next nucleotide as the misincorporated nucleotide occupies the site where the incoming nucleotide would enter.

10.2 Bacterial Transcription

10.2.1 Transcription Initiation

10.2.1.1 Key Promoter Elements in Bacteria

The bacterial RNAP core enzyme is capable of transcribing DNA into RNA in vitro. However, the core enzyme alone can only transcribe DNA templates in vitro that are nicked and not intact [12]. In vivo, the core enzyme requires an auxiliary factor, called sigma factor, to recognize sites where transcription will start and facilitate proper transcription initiation. Sigma factor associates with the core enzyme and helps RNAP not only bind to DNA but also melt the double stranded DNA so that RNAP can access the DNA template strand. Sigma factor bound to the RNAP core enzyme is called the haloenzyme and is competent to bind to specific DNA elements called promoters. Promoters in bacteria contain specific DNA elements with defined consensus sequences that are spaced a specific distance apart (Fig. 10.3).

The first DNA element within the promoter is the transcriptional start site (TSS) or the +1 site. At this site, RNAP will start transcription of the RNA molecule. Upstream of the +1, is the −10 box element which is a 6–7 bp DNA element centered 10 bp upstream of the +1 site. The region between the −10 box and the +1 is the region of DNA that is unwound

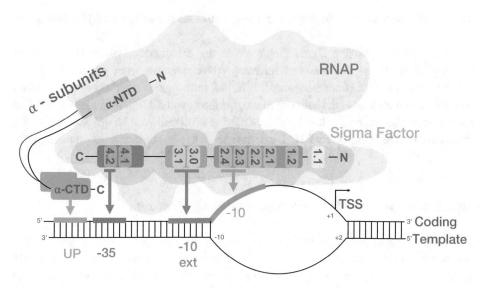

Fig. 10.3 The holoenzyme composed of RNAP (blue) and sigma factor (orange) recognize multiple elements that may define promoters in bacteria

during transcription initiation to form the transcription bubble. Though not required for promoter recognition, a −10 element with a specific 3 base pair 5′ extension (5′-TGN-3′), called the −10 extended element, can help strengthen DNA-haloenzyme interactions.

Further upstream of the +1 is the −35 box element. This element is 6 base pairs and centered approximately 35 base pairs upstream of the +1 site. The spacing between the −10 box and the −35 box elements is critical as artificial promoters, where these two elements are brought closer together, are unable to promote effective transcription initiation [13].

When multiple promoters in bacteria, such as E. coli, have been examined both the −10 box and the −35 box have strong consensus sequences. Because of this the strength of the promoter (e.g., a measure of how well transcription can occur from a promoter) is strongly correlated with how well the sequences within the −10 box and −35 box match the consensus sequence. The more each element matches the consensus sequence, the stronger the promoter will be; while promoters that deviate from the consensus sequence will weaken RNAP-promoter interactions and will initiate transcription poorly.

The strength of the promoter can also be increased by an additional element such as the −10 extended element or the UP element. The UP element is located upstream of the −35 element. Unlike the −35, −10 and the −10 extended elements, the UP element makes contacts with the RNAP core enzyme, specifically through the C-terminal domain of the α subunits (Fig. 10.3). Thus, promoters with UP elements are strongly transcribed as they more contacts between the haloenzyme and the DNA.

10.2.2 Sigma Factors Recognizes Promoters as a Part of the Haloenzyme

The core enzyme and the sigma factor create the haloenzyme, which is capable of recognizing promoter elements and initiating RNA synthesis from promoters. The σ subunit mediates interactions necessary for RNAP to recognize sites where transcription will start. Upon binding of the haloenzyme, sigma factor will aid RNAP to melt the DNA to form a transcription bubble. Interestingly, differing σ subunits can be utilized to alter the preference of which promoters RNAP can initiate on to alter gene expression patterns.

10.2.3 Recognition of the Promoter by the Haloenzyme

Transcription initiation begins with the recognition of the promoter and its elements by the haloenzyme. First, within the haloenzyme, σ recognizes and binds to the -10 and -35 elements within the promoter. When the haloenzyme (R) binds the promoter DNA (P) it forms the RNAP-promoter closed complex (RP_C, Fig. 10.4). Then σ promotes the formation of the RNAP-promoter open complex (RP_O). Here, approximately 12 base pairs of double stranded DNA near the TSS is unwound, forming a transcription bubble where the DNA template is placed into the active site. Next, a RNAP-promoter initial transcribing complex (RP_{ITC}) forms. This complex produces short RNA transcripts that are 9–11 nucleotides in length starting at the +1 site in a process called abortive transcription [14].

Abortive transcription is not productive transcription (i.e., where a full-length RNA is synthesized) as abortive RNA transcripts are not used by RNAP after they are generated and quickly disassociate from RP_{ITC}. Rather, abortive transcription is a means to generate energy to allow for RNAP to escape the promoter. During the formation of the RP_C and RP_O complexes, many interactions between RNAP and the promoter DNA are made, and for RNAP to begin productive transcription, RNAP must break those contacts so that it can translocate along the DNA template to make a full-length RNA transcript. To do this, RNAP uses the energy generated by abortive transcription to "scrunch" the DNA within itself by moving forward along the downstream DNA template while remaining stationary on the upstream DNA. This will store energy in a "stressed" intermediate, that when released will provide energy for breaking RNAP-promoter contacts, thus allowing for promoter escape [15]. RNAP performs multiple rounds of abortive transcription till enough energy is generated to escape the promoter. Upon promoter escape, the σ is released and core enzyme of RNAP becomes a competent elongation complex capable of productive transcription (RD_e).

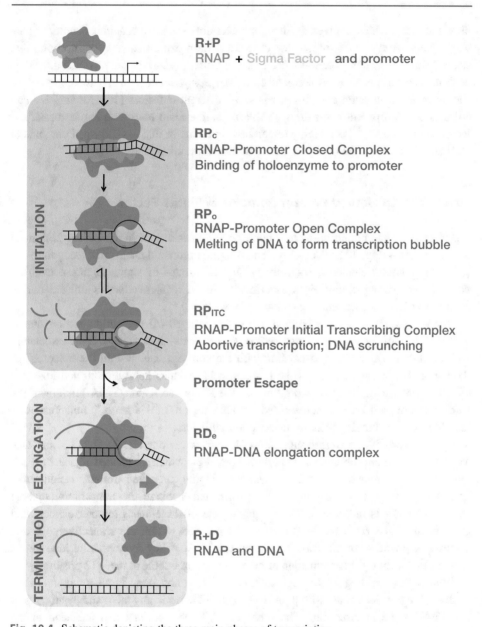

R+P
RNAP + Sigma Factor and promoter

RP_C
RNAP-Promoter Closed Complex
Binding of holoenzyme to promoter

RP_O
RNAP-Promoter Open Complex
Melting of DNA to form transcription bubble

RP_ITC
RNAP-Promoter Initial Transcribing Complex
Abortive transcription; DNA scrunching

Promoter Escape

RD_e
RNAP-DNA elongation complex

R+D
RNAP and DNA

INITIATION

ELONGATION

TERMINATION

Fig. 10.4 Schematic depicting the three main phases of transcription

10.2.4 Identity of Sigma Factors Associated with Haloenzyme Determines Gene Expression

The identity of the σ that binds to RNAP determines the set of genes that will be transcribed. Based on the environmental conditions that a bacterium encounters, several different σ factors have evolved to allow for bacterium to adapt to changing conditions they may experience by promoting a new gene expression pattern capable of surviving that stress. In one of smallest bacterial species, *Mycoplasma genitalium*, only a single σ factor is needed. A bacterium, such as the soil dwelling *Streptomyces coelicolor*, that is exposed to a variety of environmental conditions, has sixty different σ factors [16]. *Escherichia coli* contains seven different σ factors with the σ^{70} as the main housekeeping factor. The σ factor from E. coli, σ^{70}, is a highly researched sigma factor that has informed our general understanding of the role of sigma factors in transcription initiation.

10.2.5 The Conserved Protein Domains of Sigma Factor

Sigma factors can contain up to four conserved domains: domain 1, domain 2, domain 3, and domain 4 (Fig. 10.3). Domains 2 and 4 contain protein domains that recognize −10 and −35 promoter elements, respectively. Not all groups of sigma factors contain all domains. For example, domain 1 is primarily found in sigma factors utilized during conditions where housekeeping genes are expressed.

Domain 1 consists of a subregion 1.1 which is a 70–90 nucleotide section that is generally negatively charged. First, region 1.1 prevents sigma factor from associating with the DNA without first associating with the core enzyme. Within the haloenzyme, the region 1.1 of sigma contacts the β amino acid residues between 900 and 909 [16] effectively blocking the DNA from binding the active site. Upon, DNA binding of the haloenyzme, region 1.1 is then displaced from this region. In this sense, it functions as an inhibitor of DNA binding prior to transcription initiation.

Domain 2 has five conversed subregions (denoted as regions 1.2, 2.1, 2.2, 2.3, and 2.4). Most critical of these subregions are regions 2.3 and 2.4 which are located adjacent to each other on a continuous helix. Region 2.3 contains highly conserved aromatic residues that help participate in DNA melting to help form and stabilize the transcription bubble [17]. Region 2.4 is responsible for recognizing the −10 element promoter element and contacts the DNA from the −10 to the −12 position, which are respectively 10 and 12 residues upstream of the initiation site on the DNA. A key tryptophan residue in this region helps form the upstream edge of the transcription bubble at the −12 position.

Domain 3, consisting of sub-regions 3.0 and 3.1, is not found in all σ factors. Initial contact of region 3.1 occurs with β residues 1060–1240, inducing the β and β′ subunits of the RNAP to be in proximity. The subregion 3.0 helps binding and inclusion of the extended −10 promoter region only found in some promoters. This region is frequently found in bacterial promoters that do not have a −35 promoter element.

Domain 4 contains regions 4.1 and 4.2 that form an interaction with the β subunit of the core RNAP [2]. Most important is the helix-turn-helix in region 4.2 recognizes and interacts with the −35 promoter region of DNA [16]. This interaction results in a 36° bend in the DNA at the −35 element, which changes the trajectory of the upstream DNA trajectory to bring it closer to the catalytic core.

10.3   Transcription Elongation

During transcription elongation the majority of RNA synthesis occurs. In prokaryotes, transcription elongation is tightly linked to translation as the DNA and ribosomes are not found in separate cellular compartments. As the RNAP escapes the promoter and cycles through nucleotide addition to create RNA from the DNA template, RNAP will transcribe into the nascent RNA an RNA binding site, called the Shine-Dalgarno sequence, that ribosomes can bind to and begin translation from. If RNAP is transcribing an operon, there will be multiple ribosome binding sites within the RNA to denote the start of translation for each protein encoded in the operon. The ribosomes will coat the nascent RNA and perform translation as RNAP transcribes the rest of the open reading frame.

Also, during elongation, the transcription elongation complex must not only add nucleotides to synthesize RNA $5'$ to $3'$, it must also maintain the transcription bubble as it translocates along the DNA template strand. The transcription bubble is approximately 14 base pairs of melted DNA and remains this size throughout transcription elongation. This suggests that as RNAP translocates along the DNA template, it melts the upstream DNA at the same rate as the downstream DNA re-seals to form double stranded DNA, thus, keeping the transcription bubble the same size throughout elongation. To maintain the transcription bubble, RNAP forms a unique RNA-DNA hybrid with the nascent RNA and the DNA template strand. This RNA-DNA hybrid stabilizes the RD_e by preventing re-sealing the two complementary DNA strands. The $3'$ end of the nascent RNA forms a RNA-DNA hybrid with the DNA template that is approximately 8–9 base pairs in length [18]. The $5'$ end of the nascent RNA exits the elongating RNAP through an RNA exit channel near the active site. The RNA exit channel accommodates about 7 nucleotides of RNA.

While prokaryotic RNAPs are processive enzymes and elongate RNA at a rate of approximately 40 nucleotides per second [19], RNAP is found to pause transcription often during elongation. Sometimes this paused is caused by incorporating the incorrect nucleotide into the nascent RNA. In this case, the RNAP will backtrack remove the incorrect nucleotide and add the correct one. Other times, pauses in elongation to allow for the translation machinery to catch up with the elongating RNAP or allow for time to induce transcription termination if translation is not occurring.

10.4   Transcription Termination

Transcription termination primarily plays structural and regulatory roles by physically separating different parts of the genetic material. This is achieved by signaling where the end of an RNA transcript should be to the elongating RNAP. Transcription termination plays a large role in modifying gene expression patterns as prokaryotes need to coordinate rates of transcription and translation to properly express correct levels of gene products necessary for cellular function. Transcription termination is especially important because if

RNAP continues transcribing an RNA transcript, that transcript will be continually translated; thus, leading to wasteful or unnecessary gene expression.

Transcription termination signals can be classified into in one of the two groups based on their position in the genome (Fig. 10.5). The first type is found at the end of transcription units, for instance, between neighboring operons and are called intergenic. Such placement of a termination signal structurally separates different portions of the genome and permits independent expression of adjacent sequences. The second type of intragenic terminators punctuates operons into distinct genes and are considered "intra-operonic" (i.e., within an operon). Such terminators are necessary for adjusting the relative expression levels of genes within a single operon. The second group of terminators are found within genes and therefore, called intragenic. Intragenic termination signals are typically latent and turn on only under certain metabolic conditions and environmental stresses to uncouple transcription and translation.

Furthermore, there are two types of termination signals that use two distinct mechanisms to induce transcription termination (Fig. 10.5b, c). The first is called intrinsic transcription termination. Once transcribed the RNA sequence of the termination signal the will cause RNAP to disassociate without external factors. The second is called Rho mediated transcription termination. Again, a signal within the DNA is transcribed into the RNA, however, the sequence itself does not cause termination. Instead, a protein called Rho will bind to the RNA sequence, called a rut site, and then induce transcription termination.

10.4.1 Intrinsic Transcription Termination

Intrinsic terminators are found at the end of operons and distinct genes within an operon. As intragenic terminators, intrinsic terminators act as punctuation marks in the genetic material and are an important mode of regulating gene expression in prokaryotes. Additionally, many genes rely on intrinsic terminators for proper 3′ end and it is estimated that approximately 50% of annotated protein-encoding transcription units in E. coli end with an intrinsic terminator [20].

The intrinsic termination signal consists of approximately 40-nucleotide long well-defined consensus sequence with two highly invariable elements: an interrupted GC-rich dyad, followed by a 7–9 nucleotide long "A-stretch." After transcription, the GC-rich dyad of the terminator will fold into a stable secondary structure where the GC base pairs form the stem of a hairpin. The A-stretch is converted into unstructured poly(U)-tract (Fig. 10.5b) in the RNA. The stringent nucleotide sequence requirements of intrinsic terminators greatly limit the sequence of amino acids, it could encode, which possibly accounts for the rare intragenic placement of these signals.

There is currently an accepted allosteric model for intrinsic termination [21, 22]. The transcription of the region containing A-stretch causes the elongation complex to pause, thus providing time for the RNA hairpin to form within the nascent RNA. Formation of the hairpin within the RNA directly competes against the formation of the RNA-DNA hybrid

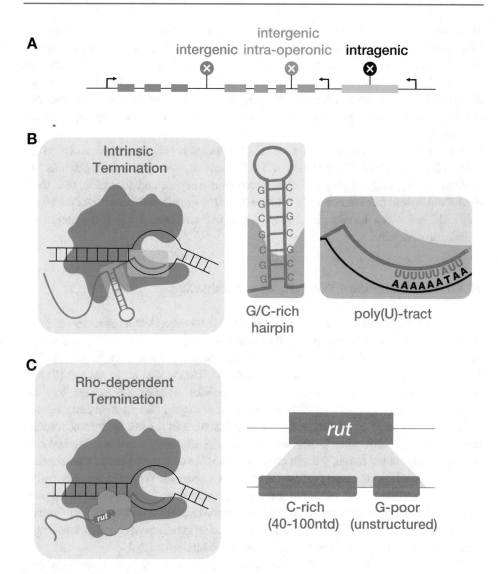

Fig. 10.5 Different types of transcription termination. A schematic depicting the location of transcription termination signals (**a**). The RNA transcript once produced can also cause termination due to structures in RNA, intrinsic termination (**b**), or by binding to rho-proteins, rho-dependent transcription termination (**c**)

near the active site of the elongating RNAP. Therefore, formation of hairpin triggers the dissociation of the DNA-RNA hybrid. At intrinsic termination sites the stability of the GC-rich RNA hairpin exceeds the stability of the AU-rich DNA-RNA hybrid as GC RNA base pairs are generally more thermodynamically stable than AU DNA-RNA base pairs.

Thus, the formation of the intrinsic terminator causes the nascent RNA to unwind from the DNA template; thereby, reducing the stability of the entire elongation complex. Furthermore, as the stem of the RNA hairpin continues to grow, it begins to invade the catalytic core of RNAP. Eventually, the hairpin reaches the trigger loop, causing it to adopt the closed conformation and making the active site inaccessible. This results in simultaneous opening of the clamp and consequent release of the nascent RNA transcript and the template DNA.

In summary, intrinsic transcription termination relies heavily on two elements in the RNA to facilitate effective termination: a G/C hairpin followed by an poly(U)-stretch. The A-stretch is essential for pausing of the elongation complex and formation of a weak RNA-DNA hybrid that is susceptible to unwinding. The growth and incursion of the highly stable GC-rich RNA hairpin destabilizes the RNAP structure, causing the elongation complex to dissociate.

10.4.2 Rho-Dependent Transcription Termination

A second mode of transcription termination occurs at Rho-dependent terminators and relies on the action of the Rho factor protein. Rho is an essential protein in *E. coli* but is absent in certain other bacterial species, such as *Staphylococcus aureus* and *Bacillus subtilis*. The Rho factor recognizes and binds to a *rut* site (*Rho Utilization site*) in the nascent RNA to induce termination approximately 60–90 nucleotides downstream of the *rut* site (Fig. 10.5c). Rho-dependent terminators are most commonly latent intragenic signals, which primarily serve regulatory and protective functions in the cell to prevent wasteful transcription of mRNA when it cannot be quickly and efficiently translated. For instance, during amino acid starvation, the rate of translation is lower than the rate of transcription, and a ribosome does not closely follow the elongation complex. When the *rut* site emerges from the elongating RNAP and is not followed by a ribosome, the Rho factor can rapidly bind the *rut* site to terminate transcription of the untranslated RNA molecules [23]. Overall, Rho-dependent termination plays a crucial role in tuning gene expression to metabolic and environmental signals by preventing unnecessary production of RNA.

The length and the sequence of characterized *rut* sites is highly degenerate but in general *rut* sites have a 40–100 nucleotide long C-rich stretch, followed by a G-poor stretch of a that lacks rigid secondary structure (Fig. 10.5c). Since the sequence of the *rut* site is highly variable, *rut* sites can be found within protein-encoding regions of the genome.

Rho has been characterized as a homohexameric ring-shaped protein with RNA-dependent ATP-dependent helicase activity. Each of the six identical protomers is about 46 kDa and contains two single-stranded RNA binding sites. These RNA binding sites recognize *rut* sites and help guide the single-stranded nascent RNA through the center cavity of the ring-shaped protein. The cavity of the hexamer only permits binding of single-stranded RNA molecules and justifies the preference for unstructured regions of RNA by Rho.

The catalytic center of the Rho factor contains RNA-dependent ATPase-helicase activity. A highly conserved Glu112 coordinates a nucleophilic water molecule to catalyze this reaction [23]. Direct contact between the nascent RNA and the Rho is required to trigger a conformational change in the Rho catalytic center that results in ATP hydrolysis. This RNA-dependent ATP hydrolysis is thought to drive translocation of Rho along the nascent RNA or give Rho the ability to unwind RNA-DNA hybrids in the elongating complex.

The mechanism by which Rho induces transcription termination is not fully understood at a fine molecular detail. However, it can be broken down into three broad events. First, Rho loads onto the *rut* site that is transcribed into the nascent RNA. Rho can only bind when there are no actively translating ribosomes occluding the *rut* site. Second, Rho translocates itself toward RNAP, however, this has been recently challenged by Mooney et al. [24] and Epshtein et al. [21] as they observe Rho associated with actively elongating RNAPs. Third, Rho induces transcription termination through its helicase activity by an undefined mechanism. Three unresolved models are suggested: (1) Rho uses helicase activity to pull apart the DNA-RNA hybrid, (2) Rho translocates into RNAP causing a conformational change that causes RNAP to disassociate without pulling on the RNA, or (3) the helicase activity of Rho pushing RNAP forward on the DNA template causing the elongating RNAP to translocate on the DNA template without NTP addition.

In summary, Rho-dependent transcription termination requires Rho factor to recognize *rut* sites within untranslated regions of a nascent RNA. Rho then induces transcription termination utilizing ATP-driven helicase activity to promote the disassembly of the elongating RNAP complex.

10.5 Eukaryotic Transcription

10.5.1 Major Differences in Eukaryotic and Prokaryotic Transcription

10.5.1.1 Eukaryotic RNAPs Are More Complex Than Bacterial or Archaeal RNAP

Eukaryotic cells contain multiple RNAPs unlike bacterial and Archaeal cells which each contain one RNAP. Although each eukaryotic RNAP differs in subunit complexity, all RNAPs have been found to have the same essential structure. Beyond this basic structure, the factors that guide RNAPs to sites of transcription differ greatly between each eukaryotic RNAP. Eukaryotic cells which contain a minimum of three different RNAPs with masses ranging from 500 to 700 kDa and each have a specific transcriptional profile. A 14-subunit RNA Polymerase I transcribes ribosomal RNA. RNA Polymerase II (RNAP II), with 12 subunits, primarily synthesizes messenger RNA and produces small regulatory RNAs. The largest, RNA Polymerase III with 17 subunits, synthesizes transfer RNA, 5S rRNA, and small noncoding RNAs. Two additional RNAPs, RNA Polymerase IV and RNA Polymerase V, have been found in plant cells, but have not yet been fully characterized. In eukaryotes, each RNAP can initiate transcription from a specific type of well-defined promoters. For example, RNAP I will initiate transcription from type I promoters, RNAPII from type II promoters and RNAP III from type III promoters. Eukaryotic RNAPs each utilizes a specific set of general transcription factors that recognize specific promoter elements to recruit RNAP and to initiate transcription.

Not only do bacteria and eukaryotic cells differ in the number of RNAPs they each have and how they initiate transcription, eukaryotic RNAP performs transcription in a double membrane bound organelle called the nucleus. This in effect partitions transcription and translation into separate compartments eliminating co-transcriptional translation in eukaryotes and adding additional levels of gene regulation. In order to transport a complete RNA transcript from the nucleus to the cytoplasmic translation machinery and protect the mRNA from degradation in the cytoplasm, mRNA transcripts (and some noncoding RNAs like those mentioned in Chap. 6) are modified at the 5′ and 3′ ends. At the 5′ end mRNAs are modified with a 7-methylguanosine cap while the 3′ end of the RNA is modified with a stretch of adenosines (called the poly(A) tail). These modifications are placed onto the RNA during transcription and help aid their effective translation by ribosomes as well as protect the RNA transcripts from degradation in the cytoplasm.

Another striking difference is the size of the eukaryotic genome versus that of the bacterial genome. The eukaryotic genome can range from ~12.5 to 1328 Mb while the largest bacterial genome is ~14.7 Mb. Because of the size and isolation of the genomic DNA within the nucleus, eukaryotic gene expression mediated by RNAPII is quite complex. DNA itself can be compartmentalized away from RNAPII to prevent its expression into biochemically defined regions called heterochromatin. Interestingly, in large eukaryotic genomes, DNA within heterochromatin is typically associated with the nuclear periphery. Some have argued that this untranscribed or "inert" DNA can shield the actively transcribed DNA from damage from ionizing sources such as UV irradiation [25]. Furthermore, a major research question in the field of gene expression has been to understand how these different domains of actively transcribed and inert DNA are established. Recent research has uncovered a role for long non-coding RNAs to either promote or limit the ability of RNAPII to access large regions of the genome. For example, a long noncoding RNAs such called HOTAIR have been demonstrated to play an integral role in partitioning large regions of inert DNA away from RNAPII [26].

10.5.1.2 CTD Domain of RNAPII and the CTD Code

The most pronounced difference among the structure of the RNAPs in found within the largest subunit of RNAP II, Rpb1. This subunit contains the carboxyterminal domain (CTD), which encompasses a critical and unique motif that allows RNAP II to be precisely regulated during transcription. This motif is highly unstructured and the repeat length of the motif—a repeat of heptapeptides—varies in different organisms. For example, the heptapeptide repeats itself 52 times in mammalian Rpb1 and 26 times in yeast, *Saccharomyces cerevisiae*. The motif generally consists of Tyr1-Ser2-Pro3-Thr4-Ser5-Pro6-Ser7, though the seventh amino acid residue is variable, with either a Lysine or an Arginine residue in place of Serine (Fig. 10.6a). Of particular interest are Ser2 and Ser5 of the CTD

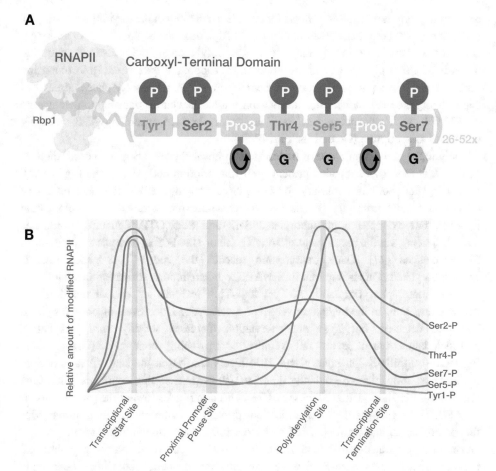

Fig. 10.6 (**a**) Schematic of the CTD of RNAPII and the post-translational modifications that can occur on the CTD repeat. (**b**) Throughout the transcription process, the CTD of RNAPII is highly susceptible to post-translational modifications on different amino acid residues in the CTD repeat such as phosphorylation, isomerization, and glycosylation. Relative amounts of modifications at different sites on RNAPII are shown for various residues; each modified residue is shown in a particular color

repeat as posttranslational modifications by kinases can phosphorylate these residues to signal distinct phases of transcription.

Throughout the transcription process, the CTD of RNAPII is highly susceptible to post-translational modifications on different amino acid residues in the CTD repeat such as phosphorylation, isomerization, and glycosylation. During early stages of transcription, modifications at Ser5 are essential. Phosphorylation of Ser5 occurs after the formation of the RNAPII open complex and is necessary to signal the transition into the elongation. On the other hand, glycosylation of Ser5 prevents phosphorylation at the same residue, thereby preventing promoter escape and allows for complete assembly of the pre-initiation complex (PIC) during transcription initiation. Once the PIC is formed, glycosylation is no longer necessary to prevent Ser5 phosphorylation. Phosphorylation at Ser5 is also critical during the early phases of elongation such as promoting 5′ capping of the nascent RNA transcript. Furthermore, when phosphorylation of Ser5 is no longer necessary, phosphatases can dephosphorylate the residue. This occurs shortly after the start of transcription elongation as RNAPII with phosphorylated Ser5 is longer detected where RNAPII normally terminates transcription (Green Line Fig. 10.6b).

Early in the elongation phase of transcription, phosphorylation begins to occur on Ser2 of the CTD repeat and remains present until transcription ends (Red Line Fig. 10.6b). Phosphorylated Ser2 is necessary for elongation, splicing, termination and export of mRNA as factors necessary for each of these actions are recruited to the elongating RNAPII. For example, a phosphorylated Ser2 and Ser5 CTD signature coordinates mRNA splicing via the recruitment of splicing factor U2AF65 to promote recognition of 3′ splice sites [27]. Once transcription reaches the end of the gene, Ser2 is dephosphorylated. This is required for RNAPII to begin another transcription cycle.

Furthermore, phosphorylation at Tyr1 and Thr4 residues are critical modifications during elongation. In the beginning stages of transcription, low levels of phosphorylated Tyr1 are detected, but Tyr1 phosphorylation increases steadily until just before polyadenylation when levels fall slightly before being eliminated (Orange Line Fig. 10.6b). Similarly, phosphorylated Thr4 levels are low at the start of transcription, but increase, though more gradually than Tyr1-P, as transcription continues (Blue Line Fig. 10.6b). Phosphorylation at these two residues contributes toward the processivity of RNAPII. These post-translation modifications prevent premature termination of transcription by preventing the binding of factors necessary for transcription termination.

Another post-translational modification of the CTD can occur on the peptide bond between Ser2-Pro3 or Ser5-Pro6 which can undergo isomerization where the peptide bond can be rotated from *cis* to *trans* or *trans* to *cis* (Fig. 10.6a). The effect on transcription is dependent on whether the proline residues are in a *cis* versus *trans* orientation after isomerization, thus allowing for different factors to bind to the CTD. In yeast, proline isomerization has been shown to play multiple roles during transcription. Peptidyl-prolyl isomerases (PPIases) that recognize phosphorylated serine residues and rotate the Serine-Proline peptide bond aid transcription termination by promoting the dephosphorylation of Ser5 [28]. Without the yeast phosphorlyated-Ser5-Pro6 PPIase, Ess1, transcription does not terminate properly and leads to the formation of run-on transcripts (RNA transcripts that extend beyond the normal termination site) [29]. Additional transcriptional defects are found in other stages of transcription when PPIases that on peptide bonds in the CTD RNAPII are inhibited. These observed effects alter transcription initiation at inducible genes and transcription elongation.

Other posttranslation modifications to the CTD of RNAPII have been discovered including Ser7 phosphorylation, glycosylation and methylation; with many more likely to be discovered. Many of these posttranslational modifications have been associated with

RNAPII in a specific spatiotemporal space or stage of transcription. For example, phosphorylated Ser5 is associated with initiating RNAPII at the 5′ end of genes (Green, Fig. 10.6b) while phosphorylated Ser2 is associated with elongating RNAPII at the 3′ end of genes (Red, Fig. 10.6b). The spatiotemporal association of posttranslational modifications to the CTD has many in the field proposing that a "CTD code" exists. In the CTD code hypothesis, it is postulated that a specific pattern of post-translational modifications are placed on the CTD of RNAPII to help recruit a specific set of factors that are necessary at that time to coordinate the various events during transcription. This CTD code thereby allows for specific transcriptional events to occur in a correct temporal manner during transcription. Most interestingly, a CTD code may underlie the coordination of transcription and RNA processing, such as splicing and 5′ capping, to occur co-transcriptionally.

10.5.2 The Core RNAP II Promoter

To recruit RNAP to regions of DNA that will be transcribed into RNA, each eukaryotic polymerase has a specific set of general transcription factors that will recognize specific DNA elements. These general transcription factors along with the RNAP assemble a Pre-Initiation Complex (PIC) competent for transcription initiation. In particular, the assembly of the RNAPII PIC on promoters occurs on a minimal DNA sequence termed a core promoter. Core promoters can facilitate transcription initiation, however, a core promoter on its own functions at a low efficiency and does not stimulate high levels of transcription. Transcriptional activators, which function to stimulate transcription from core promoters, act to either strengthen PIC interactions, mediate PIC assembly, or increase the accessibility of the DNA binding sites within the promoter.

Initiation of RNAPII transcription occurs on core promoters which contain specific DNA elements that allow for the correct assembly and orientation of the PIC. Core RNAPII promoters typically contain elements that extend approximately 35–40 base pairs upstream and/or downstream of the TSS. Seven core promoter elements have been identified and characterized [30]. A majority of RNAPII promoters contain one or more of these elements. However, it is rare that all these elements are observed within a single promoter.

The most common DNA element found within core promoters is the initiator (*Inr*) element. It is estimated to be present in ~50% of human core promoters. The *Inr* has the consensus sequence $(C/G/T)_2CA(C/G/T)(A/T)$ and centered on the TSS. The A within the *Inr* consensus sequence is the transcription start site. Studies from the Baltimore lab have shown that the *Inr* sequence alone can form a functional promoter but can be stimulated greatly by other core promoter elements [31].

Some RNAPII core promoters contain a TATA box element and are the most well characterized examples of core promoter elements. Core promoters that contain TATA box elements are characterized by an A/T rich region located approximately 25–30 nucleotides upstream of the TSS in humans. The TATA box seems to be required for one of two functions: to position where transcription will start or to initiate RNA synthesis. In some promoters, removal of the TATA box causes transcription to try to initiate at multiple sites while in other genes; causing transcription to never initiate. The TATA box is recognized specifically by the TATA-binding protein (TBP), which is a subunit of RNAPII general transcription factor TFIID. Initially, the TATA box was thought of as a common element in all core promoters as TBP is universally required for RNAPII transcription. However, bioinformatics analysis has shown a functional TATA box is only utilized in one-third of human gene promoters [30]. This evidence suggests that many RNAPII promoters require TBP/TFIID function but do not recruit TBP/TFIID through TATA box elements.

Promoters without TATA boxes are called TATA-less promoters. Often, TATA-less promoters contain other DNA elements within the core promoter to allow for transcription initiation and recruitment of TBP/TFIID. One example is the downstream promoters element (DPE) which is found downstream of the TSS and have a consensus sequence of G(AT/T)CG. In *Drosophila,* most TATA-less promoters that recruit TBP/TFIID contain a *Inr* and a DPE element. Together, these two elements successfully recruit TBP/TFIID through TBP-Associated Factors (TAFs). TAFs are auxiliary subunits of TFIID and 13 TAFs associate with TBP to form the TFIID complex. In the case of TATA-less promoters that contain *Inr* and DPE elements, TAF1 and TAF2 are required for TBP recruitment where TAF1 and TAF2 make synergistic contacts with *Inr* while DPE elements seem to strengthen TAF1/2 interactions with *Inr* by contacting TAF6 and TAF9.

The motif-ten-element (MTE), is another core promoter element found to enhance *Inr*-TAF1-TAF2 contacts in TATA-less promoters, like DPE, through associating with TAF6 and TAF9. MTE elements are thought to cooperate with the initiator to stimulate transcription in a manner independent to DPE action. MTE elements are generally found upstream of DPE where most MTE-containing promoters also contain a DPE element.

Another set of prominent core promoter elements are TFIIB-recognition elements called BRE^u and BRE^d which lie upstream and downstream of the TATA box, respectively. These elements are recognized by the helix-turn-helix motif of TFIIB where binding of TFIIB to BRE^u establishes a favorable orientation for further PIC assembly. The recognition loop of TFIIB contains amino acid residues that make contact with BRE^d, further generating more contacts between TFIIB and the DNA two stabilize the TFIIB-TBP-promoter DNA complex. BRE^u and BRE^d are found in both TATA-containing and TATA-less promoters to increases the strength of the core promoter.

Though the core promoter and its elements only dictate basal levels of transcription, the strength of the core promoter can greatly differ based on composition of the elements found within that specific core promoter. For example, synthetic core promoter sequences have been made containing *Inr*, MTE, DPE, and a TATA box. These artificial promoters with multiple elements have been demonstrated to be stronger than natural core promoters. Therefore, the level of gene expression, in part, is modulated by the combination of elements found within the core promoter.

10.5.3 Recognition of RNAPII Promoter by GTFs

10.5.3.1 General Mechanism of PIC Assembly

A multi-protein complex consisting of RNAPII and general transcription factors (GTFs) are the minimal elements required to start the process of transcription. This assembly of transcription machinery is called the pre-initiation complex (PIC) and summarized in Fig. 10.7. The GTFs of RNAPII consist of TFIIA, TFIIB, TFIID, TFIIE, TFIIF and TFIIH, all of which are relatively well conserved among eukaryotes. Biochemical and structural studies involving RNAPII and the GTFs have shed light on the assembly process of the PIC.

To understand the assembly mechanism of the PIC, early in vitro transcription experiments were performed with purified transcription factors and RNAPII. These results supported a sequential assembly model to produce a stable PIC product [30]. The sequential assembly model was: (1) the binding of the TATA box by the TBP subunit of TFIID; (2) stabilization of the TFIID/promoter complex by binding of TFIIA and TFIIB; (3) recruitment of TFIIF/RNA polymerase complex; (4) then binding of TFIIE and TFIIH (Fig. 10.7).

An alternative assembly model was proposed after a holoenzyme complex, not bound to the core promoter, was isolated. The holoenzyme used in this study was a preassembled complex of RNAPII and all of the GTFs, except for TFIID and TFIIA, indicating that RNAPII and a majority of the GTFs might preassemble before making contact with the promoter. In this model, TFIID, acts as a core promoter-binding factor, and is able to recruit a stable, preassembled RNAPII holoenzyme to the promoter region, completing the PIC assembly in two steps rather than the four described above [30].

These two assembly models proposed above have recently been challenged by the analysis of PIC assembly using the fluorescence recovery after photobleaching (FRAP) method [32]. FRAP can determine binding times and diffusion coefficients of the GTFs by labeling the GTFs and RNAPII with green fluorescent proteins (GFPs), which is followed by photobleaching the protein bound GFPs and monitoring the fluorescence recovery time. If the transcription machinery were stable, each GTF would have a relatively slow FRAP recovery. The results of the FRAP experiments favor a more dynamic assembly model, arguing against stable complexes. Interestingly, the FRAP data estimates that only 1 in 90 polymerases proceed to elongation [33]. This suggests that the GTFs and RNAPII assemble into unstable PIC complexes, opening the possibility that PIC assembly may occur stochastically, instead of sequentially as previously proposed.

10.5.3.2 Promoter Recognition Followed by Closed Complex Formation

Once a promoter is recognized and PIC assembly occurs, the PIC will prepare to initiate transcription by forming a closed complex formation. For TATA box-containing promoters, the C-terminal of TBP is able to recognize the eight base pairs in the minor groove of the TATA element in a directional manner. In humans, TBP uses two pairs of phenylalanine residues to induce an 80°–90° bend in the DNA. Intercalation of Phe284/

Fig. 10.7 Eukaryotic RNAPII
promoter recognition is
mediated by RNAPII general
transcription factors

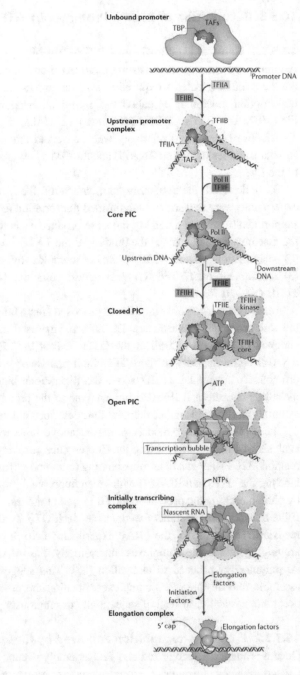

Nature Reviews | Molecular Cell Biology

Phe301 at the 5′ end and Phe193/Phe210 at the 3′ end of the TATA element causes DNA bending [34] (Fig. 10.7). The TBP-TATA box minor groove interface is dominated by van der Waals interactions. The presence of TFIIA can enhance TBP binding through direct contacts with TBP and the DNA sequence upstream of the TATA box.

After binding of TFIID and TFIIA, the C-terminal of TFIIB binds to TBP, BRE^u and BRE^d, acting as a scaffold for RNAPII and the remaining GTFs. The N-terminal domains of TFIIB, called the zinc ribbon and B-finger, can recognize and interact with the RNAPII surface, helping it enter the assembly in its correct orientation. The zinc ribbon binds to the RNAPII subunits Rpb1 and Rpb2. The B-finger enters the active site, essentially creating a tunnel for the DNA template strand to enter the active site. The domains of TFIIB, thus, can guide DNA downstream of the TATA box near the RNAPII active site. This closed complex has the DNA helix properly positioned to form the RNAP open complex.

10.5.3.3 Formation of the RNAPII Open Complex

Like bacterial RNAP, the RNAPII closed complex transitions to an open complex, where DNA is melted to form a transcription bubble; thus, allowing RNAPII to correctly position itself on around the transcriptional start site.

Establishment of the transcription bubble is carried out by TFIIH. TFIIH possesses a DNA-dependent ATPase-helicase, p89/Xeroderma pigmentosum complementation group B (XPB), which has been identified as the catalytic subunit of TFIIH required for open complex formation. However, the mechanism by which p89/XPB facilitates open complex formation remains unresolved. In the first model p89/XPB is thought to bind to a tightly wound DNA helix bound by GTFs and RNAPII. This tightly wound DNA helix is thought to be unstable and ATPase activity of p89/XPB has been suggested to pull on exposed ssDNA near the TSS, opening the DNA helix. Crosslinking studies support this model as p89/XPB makes promoter contacts both downstream and upstream of the TSS but fail to make the same contacts in mutants of TFIIF that do not form tightly wound DNA-RNAP contacts and/or allow for open complex formation [35]. A later crosslinking study shows p89/XPB only interacts with promoter DNA downstream of the TSS [36]. The distance of p89/XPB from where the transcriptional bubble would form suggests that the ATP-dependent helicase activity of p89/XBP establishes the open complex formation by acting as a "molecular wrench." In this mechanism, ATP hydrolysis induces negative helical torsion by rotating downstream promoter DNA. The torsion causes DNA helix separation because the DNA upstream of XPB activity is rotationally fixed due to promoter contacts of the TBP-TFIIB-RNAPII complex. Recent structural studies of TFIIH also show that p89/XPB only contacts promoter DNA downstream of the TSS [37], further supporting the "molecular wrench" mechanism for open complex formation.

TFIIH/XPB works in tandem with TFIIB, TFIIE, and TFIIF, to stabilize the unfavorable separation of a DNA helix. It is believed that TFIIF and TFIIE contain residues that stabilize the coding strand after the DNA has been melted. This, in turn, will increase stability and flexibility of the template strand so it can be pulled down into tunnel toward the active site by the N-terminal domains of TFIIB, allowing the formation of the transcription bubble from −9 to +2 sites in the promoter [38].

10.5.3.4 RNA Polymerase II Promoter Escape

Strong promoter contacts are maintained by the GTFs throughout the initial synthesis of nascent RNA. The strong protein–DNA interactions must be broken for the RNAPII to escape the promoter and make the transition from transcription initiation into elongation. The DNA-protein interactions are broken by PIC instability that is induced by the transcription bubble, TFIIH and TFIIB. Once the instability of the PIC has reached a critical threshold, the PIC will undergo a conformational change allowing it to release itself from its promoter contacts.

During initial synthesis of the nascent RNA, the transcription bubble remains fixed at its upstream edge while it extends downstream due to the helicase activity of TFIIH/XPB. TFIIH/XPB is hypothesized to slide along DNA ahead of, but still attached to RNAPII, extending the transcription bubble by using its helicase activity to unwind the DNA helix [39]. The stability of PIC decreases due to the weakening of promoter contacts caused by TFIIH/XPB progressing downstream to extend the bubble. Extension of the transcription bubble to approximately 18 base pairs creates almost enough instability for the PIC to disrupt its promoter contacts. TFIIB acts to further increase PIC instability.

Since the B-finger of TFIIB resides close to the DNA template strand near the RNAPII active site, it is in the direct path of the advancing 5′ end of the nascent RNA. Therefore, the B-finger can help stabilize the initially short, weak nascent RNA strand made during abortive transcription that is required for promoter clearance. Once the nascent RNA strand is synthesized to a length beyond 5 or 6 nucleotides, the B-finger must compete with the nascent RNA strand for space within RNAPII. A nascent RNA strand longer than 5 or 6 nucleotides can displace the B-finger, and thus, further reducing stability of the PIC. The transcription bubble continues to open and reach about 18 melted base pairs in size. At the same time the nascent RNA strand is extended beyond 5 or 6 nucleotides. These events are associated to another transition during promoter escape where 8 base pairs from the fixed region on the upstream edge of the transcription bubble are reannealed [40]. This is referred to as the bubble collapse and marks the last step of eukaryotic transcription initiation.

10.6   Transcription Elongation

The elongating RNAPII is highly processive enzyme like the bacterial RNAP. However, unlike the bacterial RNAP, elongating RNAPII has to coordinate RNA processing events that occur co-transcriptionally such as 5′ capping, splicing (detailed in Chap. 6) and polyadenylation. Many of these events are coordinated by loading factors necessary for RNA processing on the CTD of RNAPII where many recognize and bind to a specific posttranslationally modified form of CTD (e.g., phosphorylated Ser5). Furthermore, co-transcriptional RNA processing seems to be sensitive to the speed at which RNAPII elongation occurs as fast or slow elongation impacts promote specific splicing or polyadenylation patterns of RNA transcripts. Below, we discuss recent research in transcription elongation events that coordinate RNA processing.

10.6.1 Proximal Promoter Pausing

Early on during transcription elongation, RNAPII is observed to pause near the promoter shortly after transcription initiation. Proximal promoter pausing seems biologically relevantly as the length of the pause can act as point of regulation and a time at which 5′ capping of the mRNA can occur in a coordinated manner. In *Drosophila*, this pause occurs 20–60 base pairs from the TSS [41]. Interestingly, the amount of RNAPII found paused in the proximal promoter region can vary from gene to gene. For example, genes that are induced by signaling pathways tend to have more paused RNAPII in the proximal promoter region relative to other genes [42]. This finding suggests that proximal promoter pausing of RNAPII may be another regulatory point where gene expression can be regulated. This form of transcriptional regulation is of great interest as the cell may prime RNAPII for rapid RNA expression, thus, allowing for a quicker cellular response to an external signal/cue. Gene expression can rapidly occur as RNAPII is already initiated and waiting for release from the pause to continue transcription elongation.

Proximal promoter pausing has also been shown to facilitate 5′ capping of the nascent RNA and help promote the proper export and translation of the subsequent mRNA. During the pause, the Cap Binding Complex (CBC) and the capping enzymes are recruited to the CTD of RNAPII as the binding of these complexes to RNAPII are stimulated by phosphorylation of Ser5 of the CTD repeat. The capping enzymes cap the RNA transcript with a 7-methylguanosine while the CBC will bind to the newly placed 5′ cap, remaining associated with the cap till the mRNA is translated. mRNA bound by CBC is an important marker for the cell as it signals that the mRNA is competent for both nuclear export and translation. Once the cap is placed, a Ser2 CTD kinase, *p*-TEFb, is recruited to the pause RNAPII and phosphorylates Ser2 of the CTD repeat, releasing RNAPII from the pause and allowing it to continue transcription.

10.6.2 Kinetic Coupling of the Rate of Elongation and RNA Processing

During transcription elongation, other co-transcriptional RNA processing events occur such as splicing and polyadenylation. Recently, it has been proposed that the rate of elongation can impact these two RNA processing events [43]. Both splicing and polyadenylation can have multiple sites within a gene that can be utilized. Based on which splice sites are used for a particular mRNA transcript, the coding sequence can be altered to produce different mRNA isoforms that encode for different proteins from the same gene. Also, usage of alternative polyadenylation sites can add or remove RNA binding sites in the 3′ untranslated region of the mRNA. These binding sites can alter the amount of expression of the resulting mRNA by affecting their ability to be translated, their subcellular localization, or rate of degradation in the cell. Lastly, splice sites and/or polyadenylation sites that are utilized are marked after RNA processing to indicate that the mRNA is mature. These marks are recognized by nuclear export factors to facilitate export of mature mRNA.

Sites where splicing and polyadenylation occur at are defined by RNA elements each with a specific consensus sequence, however, not all splice sites and polyadenylation sites utilized match their respective consensus sequence. This means that there are strong (i.e., matches the consensus sequence perfectly) and weak (i.e., does not perfectly match the consensus sequence) sites where splicing or polyadenylation can occur. A proposed "window of opportunity" model has emerged and postulates that splice site and polyadenylation site choice is influenced by the rate of elongation. In this model (Fig. 10.8), if a weak site is upstream of a strong site, the speed of RNAPII elongation will affect which site is utilized for RNA processing. In this scenario, if the elongating RNAPII transcribes the region near the upstream weak site slowly, there is a longer "window" for the weaker site to be utilized as the strong downstream site is not yet transcribed into the RNA and cannot be recognized by RNA processing factors. However, if the RNAPII is elongating at a faster rate, that "window" to use the weaker site is shortened as the strong site will appear in the mRNA in a shorter amount of time. The presence of the stronger site will then outcompete the splicing or polyadenylation factors from the weak site; thus, preferencing an mRNA isoform generated from the use of stronger sites for processing.

10.7   Transcriptional Termination

While transcription initiation has been highly studied, how RNAPII transcription terminates is a relative new field of study. Unlike prokaryotes, eukaryotic RNAs transcribed by RNAPII are processed during transcription to modify the RNA such as the addition of a 5′ cap, removal of introns via splicing, and 3′ polyadenylation. The latter is tightly coupled with transcription termination.

The most studied mechanism of transcription termination consists of two steps. First, the 3′ end of RNA must be defined; and second, the template DNA and nascent RNA must be released from the elongating RNAPII. In eukaryotes, the poly(A) signal (PAS) acts as a transcription termination site (TTS). In humans, the PAS signal consists of the AU-rich nucleotide sequence AAUAAA. As soon as the PAS is transcribed, 3′ end processing factors interact with the mRNA, some of which associate with the CTD of Rpb1 once Ser2 has been phosphorylated. With the recruitment of the factors required for 3′ end processing subsequent endonucleolytic cleavage and polyadenylation of the RNA occurs. Therefore, cleavage and polyadenylation essentially releases an RNA competent for nuclear export from the elongating RNAPII. The RNA transcript that remains associated with RNAPII

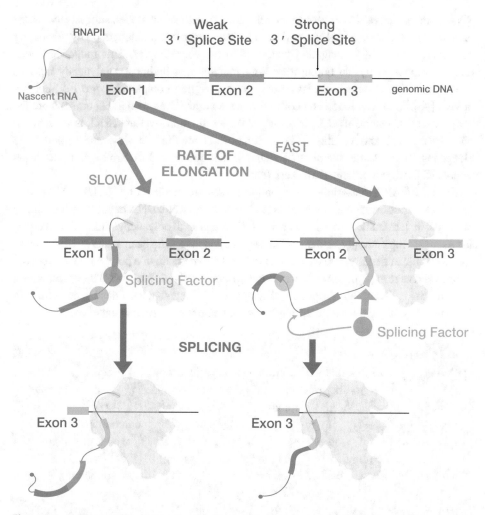

Fig. 10.8 A diagram illustrating the window of opportunity model for co-transcriptional RNA processing

after cleavage is unprotected as it is not capped on 5′ end. In yeast, it is believed that RNAPII continues transcription no more than 200 base pairs downstream of cleavage and polyadenylation [44]. Currently there are two extensively studied mechanisms to disengage RNAPII and terminate transcription after cleavage and polyadenylation.

Interacting with the CTD of RNAPII is an exonuclease called Rat1/Xrn2. This exonuclease is loaded onto the 5′ end of the RNA transcript emanating from RNAP after cleavage. A long-favored model for Rat1/Xrn2-mediated transcription termination is the Torpedo Model. In this model, Rat1/Xrn2 is thought to work its way toward the elongating RNAPII, eventually colliding with RNAPII to stop transcription and promote release of the DNA by degrading the RNA through its conserved exonuclease activity or by removing the

RNA from the active site. A second model, called the allosteric model, suggests that Rat1/Xrn2 help remodel the RNAPII complex to be less competent for elongation by preventing or removing necessary elongation factors from the core enzyme. This model has been recently supported by the finding that Rat1 mutants that harbor a catalytically inactive exonuclease domain can still disassembly a transcription complex in vitro [45] but not in vivo [46]. Recently, Eaton and colleagues have argued for a model that unifies both the Torpedo and allosteric model for transcription termination by Rat1/Xrn2 as they find a phosphatase that acts on the CTD of RNAPII is required to slow and remodel the elongating RNAPII after the poly(A) signal to help promote Rat1/Xrn2 activity to disassemble RNAPII and terminate transcription.

Not all RNAPII transcription termination events are mediated by Rat1/Xrn2. Another pathway for transcription termination is facilitated by an RNA/DNA helicase called Sen1 is involved in terminating transcription after cleavage and polyadenylation of noncoding RNAs [47]. Interestingly, transcription termination has also been observed in a small subset of human genes where cleavage is induced downstream of a poly(A) site and occurs prior to cleavage and polyadenylation of the upstream poly(A) site [48]. These termination elements are called co-transcriptional cleavage (CoTC) elements and have been defined in ~80 human genes. However, their mechanism of action still remains undefined.

Take Home Message

RNAP plays a central role in transcribing the genetic information encoded within the DNA. The multi-protein enzyme is highly conserved from prokaryotes, archaeal and eukaryotes to synthesize RNA through three defined phases of transcription. RNAP first identifies and binds to promoter DNA during transcription initiation. The DNA is subsequently unwound to expose the template strand and engage the DNA within its active site. During transcription elongation RNAP utilizes rNTPs and the DNA template within its active site to generate an RNA molecule through the nucleotide addition cycle in a highly processive manner. Finally, at the end of a gene, transcription is signaled to end and RNAP dissociates from the DNA and nascent RNA in the last step of transcription, called transcription termination.

References

1. Vassylyev DG, Sekine S, Laptenko O, et al. Crystal structure of a bacterial RNA polymerase holoenzyme at 2.6 a resolution. Nature. 2002;417:712–9.
2. Murakami KS, Masuda S, Campbell EA, et al. Structural basis of transcription initiation: an RNA polymerase holoenzyme-DNA complex. Science. 2002;296:1285–90.
3. Gnatt AL, Cramer P, Fu J, et al. Structural basis of transcription: an RNA polymerase II elongation complex at 3.3 Å resolution. Science. 2001;292:1876–82.
4. Wang D, Bushnell DA, Westover KD, et al. Structural basis of transcription: role of the trigger loop in substrate specificity and catalysis. Cell. 2006;127:941–54.

5. Westover KD, Bushnell DA, Kornberg RD. Structural basis of transcription: nucleotide selection by rotation in the RNA polymerase II active center. Cell. 2004;119:481–9.

6. Todone F, Brick P, Werner F, et al. Structure of an archaeal homolog of the eukaryotic RNA polymerase II RPB4/RPB7 complex. Mol Cell. 2001;8:1137–43.

7. Mazumder A, Lin M, Kapanidis AN, Ebright RH. Closing and opening of the RNA polymerase trigger loop. Proc Natl Acad Sci. 2020;117:15642–9.

8. Sydow JF, Brueckner F, Cheung ACM, et al. Structural basis of transcription: mismatch-specific Fidelity mechanisms and paused RNA polymerase II with frayed RNA. Mol Cell. 2009;34:710–21.

9. Belogurov GA, Artsimovitch I. The mechanisms of substrate selection, catalysis, and translocation by the elongating RNA polymerase. J Mol Biol. 2019;431:3975–4006.

10. Kuznedelov K, Korzheva N, Mustaev A, Severinov K. Structure-based analysis of RNA polymerase function: the largest subunit's rudder contributes critically to elongation complex stability and is not involved in the maintenance of RNA–DNA hybrid length. EMBO J. 2002;21:1369.

11. Cheung ACM, Cramer P. Structural basis of RNA polymerase II backtracking, arrest and reactivation. Nature. 2011;471:249–53.

12. Burgess RR, Travers AA, Dunn JJ, Bautz EKF. Factor stimulating transcription by RNA polymerase. Nature. 1969;221:43–6.

13. Dombroski AJ, Johnson BD, Lonetto M, Gross CA. The sigma subunit of Escherichia coli RNA polymerase senses promoter spacing. Proc Natl Acad Sci U S A. 1996;93:8858–62.

14. Alhadid Y, Chung S, Lerner E, et al. Studying transcription initiation by RNA polymerase with diffusion-based single-molecule fluorescence. Protein Sci. 2017;26:1278–90.

15. Kapanidis AN, Margeat E, Ho SO, et al. Initial transcription by RNA polymerase proceeds through a DNA-scrunching mechanism. Science. 2006;314:1144–7.

16. Gruber TM, Gross CA. Multiple sigma subunits and the partitioning of bacterial transcription space. Annu Rev Microbiol. 2003;57:441–66.

17. Campbell EA, Muzzin O, Chlenov M, et al. Structure of the bacterial RNA polymerase promoter specificity sigma subunit. Mol Cell. 2002;9:527–39.

18. Nudler E, Mustaev A, Goldfarb A, Lukhtanov E. The RNA–DNA hybrid maintains the register of transcription by preventing backtracking of RNA polymerase. Cell. 1997;89:33–41.

19. Proshkin S, Rahmouni R, Mironov A, Nudler E. Cooperation between translating ribosomes and RNA polymerase in transcription elongation. Science. 2010;328:504–8.

20. Lesnik EA, Sampath R, Levene HB, et al. Prediction of rho-independent transcriptional terminators in Escherichia coli. Nucleic Acids Res. 2001;29:3583–94.

21. Epshtein V, Cardinale CJ, Ruckenstein AE, et al. An allosteric path to transcription termination. Mol Cell. 2007;28:991–1001.

22. Gusarov I, Nudler E. The mechanism of intrinsic transcription termination. Mol Cell. 1999;3:495–504.

23. Banerjee S, Chalissery J, Bandey I, Sen R. Rho-dependent transcription termination: more questions than answers. J Microbiol. 2006;44:11–22.

24. Mooney RA, Davis SE, Peters JM, et al. Regulator trafficking on bacterial transcription units in vivo. Mol Cell. 2009;33:97–108.

25. García-Nieto PE, Schwartz EK, King DA, et al. Carcinogen susceptibility is regulated by genome architecture and predicts cancer mutagenesis. EMBO J. 2017;36:2829–43.

26. Tsai M-C, Manor O, Wan Y, et al. Long noncoding RNA as modular scaffold of histone modification complexes. Science. 2010;329:689–93.

27. David CJ, Boyne AR, Millhouse SR, Manley JL. The RNA polymerase II C-terminal domain promotes splicing activation through recruitment of a U2AF65–Prp19 complex. Genes Dev. 2011;25:972–83.

28. Egloff S, Zaborowska J, Laitem C, et al. Ser7 phosphorylation of the CTD recruits the RPAP2 Ser5 phosphatase to snRNA genes. Mol Cell. 2012;45:111–22.
29. Ganem C, Devaux F, Torchet C, et al. Ssu72 is a phosphatase essential for transcription termination of snoRNAs and specific mRNAs in yeast. EMBO J. 2003;22:1588–98.
30. Thomas MC, Chiang C-M. The general transcription machinery and general cofactors. Crit Rev Biochem Mol Biol. 2006;41:105–78.
31. Smale ST, Baltimore D. The "initiator" as a transcription control element. Cell. 1989;57:103–13.
32. Stasevich TJ, McNally JG. Assembly of the transcription machinery: ordered and stable, random and dynamic, or both? Chromosoma. 2011;120:533–45.
33. Sprouse RO, Karpova TS, Mueller F, et al. Regulation of TATA-binding protein dynamics in living yeast cells. Proc Natl Acad Sci. 2008;105:13304–8.
34. Nikolov DB, Chen H, Halay ED, et al. Crystal structure of a human TATA box-binding protein/ TATA element complex. Proc Natl Acad Sci. 1996;93:4862–7.
35. Douziech M, Coin F, Chipoulet J-M, et al. Mechanism of promoter melting by the Xeroderma Pigmentosum complementation group B helicase of transcription factor IIH revealed by protein-DNA photo-cross-linking. Mol Cell Biol. 2000;20:8168–77.
36. Spangler L, Wang X, Conaway JW, et al. TFIIH action in transcription initiation and promoter escape requires distinct regions of downstream promoter DNA. Proc Natl Acad Sci. 2001;98: 5544–9.
37. Gibbons BJ, Brignole EJ, Azubel M, et al. Subunit architecture of general transcription factor TFIIH. Proc Natl Acad Sci. 2012;109:1949–54.
38. Bushnell DA, Westover KD, Davis RE, Kornberg RD. Structural basis of transcription: an RNA polymerase II-TFIIB Cocrystal at 4.5 Angstroms. Science. 2004;303:983–8.
39. Wang X, Spangler L, Dvir A. Promoter escape by RNA polymerase II: downstream promoter DNA is required during multiple steps of early transcription. J Biol Chem. 2003;278:10250–6.
40. Pal M, Ponticelli AS, Luse DS. The role of the transcription bubble and TFIIB in promoter clearance by RNA polymerase II. Mol Cell. 2005;19:101–10.
41. Nechaev S, Fargo DC, dos Santos G, et al. Global analysis of short RNAs reveals widespread promoter-proximal stalling and arrest of Pol II in Drosophila. Science. 2010;327:335–8.
42. Adelman K, Lis JT. Promoter-proximal pausing of RNA polymerase II: emerging roles in metazoans. Nat Rev Genet. 2012;13:720–31.
43. Bentley DL. Coupling mRNA processing with transcription in time and space. Nat Rev Genet. 2014;15:163–75.
44. Birse CE, Minvielle-Sebastia L, Lee BA, et al. Coupling termination of transcription to messenger RNA maturation in yeast. Science. 1998;280:298–301.
45. Dengl S, Cramer P. Torpedo nuclease Rat1 is insufficient to terminate RNA polymerase II in vitro. J Biol Chem. 2009;284:21270–9.
46. Kim M, Krogan NJ, Vasiljeva L, et al. The yeast Rat1 exonuclease promotes transcription termination by RNA polymerase II. Nature. 2004;432:517–22.
47. Nedea E, Nalbant D, Xia D, et al. The Glc7 phosphatase subunit of the cleavage and polyadenylation factor is essential for transcription termination on snoRNA genes. Mol Cell. 2008;29:577–87.
48. Nojima T, Dienstbier M, Murphy S, et al. Definition of RNA polymerase II CoTC terminator elements in the human genome. Cell Rep. 2013;3:1080–92.

Index

© Springer Nature Switzerland AG 2022
N. Grover (ed.), *Fundamentals of RNA Structure and Function*, Learning Materials in Biosciences, https://doi.org/10.1007/978-3-030-90214-8

Printed in the United States
by Baker & Taylor Publisher Services